AN OLD MAN'S
TOY

AN OLD MAN'S TOY

Gravity at Work and Play in Einstein's Universe

A. Zee

COLLIER BOOKS
Macmillan Publishing Company • New York

Collier Macmillan Canada • Toronto

Maxwell Macmillan International
New York   Oxford   Singapore   Sydney

Collier Books
Macmillan Publishing Company
866 Third Avenue, New York, NY 10022

Collier Macmillan Canada, Inc.
1200 Eglinton Avenue East, Suite 200
Don Mills, Ontario M3C 3N1

Lines from "Begin the Beguine" (page 120), words and music by
Cole Porter © 1935 by Warner Bros. Inc. (Renewed). All rights reserved.
Used by permission.

Lines from "Fire and Ice" by Robert Frost (page 207) copyright 1923
by Holt, Rinehart and Winston and renewed 1951 by Robert Frost. Reprinted
from *The Poetry of Robert Frost*, edited by Edward Connery Lathem,
by permission of Henry Holt and Company, Inc.

Library of Congress Cataloging-in-Publication Data
Zee, A.
    An old man's toy : gravity at work and play in Einstein's universe /
A. Zee.— 1st Collier Books ed.
        p.   cm.
    Includes bibliographical references (p.   ) and index.
    ISBN 0-02-040915-X
    1. Gravitation—Popular works.   2. Gravity—Popular works.
    3. Cosmology—Popular works.   I. Title.
    [QC178.Z44   1990]
    531'.14—dc20   90-2328   CIP

Macmillan books are available at special discounts for bulk purchases for sales promotions, premiums, fund-raising, or educational use. For details, contact:

    Special Sales Director
    Macmillan Publishing Company
    866 Third Avenue
    New York, NY 10022

First Collier Books Edition 1990

10 9 8 7 6 5 4 3 2 1

Printed in the United States of America

To Chubber & Droid

Contents

Contents

Preface

Of the fundamental forces of nature, we are most intimate with gravity. In the uttermost darkness of night, lost in our private thoughts and shut off from the world of light, we still feel the incessant tug of gravity. No sooner had we come into existence than we became aware of the downward pull of gravity, balanced by the buoyancy of the fluid inside our mothers' wombs. Yet we do not know gravity.

Physics began with gravity. Two great theorists, Newton and Einstein, one building upon the other, strove to reveal the secrets of this universal force. Since Einstein, physicists have pondered, argued, and fought about the underlying character of gravity. The force we are most intimate with is a mysterious interloper from a realm of energy far beyond our experience. We know gravity by happenstance.

The titanic clash between gravity and the quantum has rocked the physics world for half a century. In recent years, a strangely beautiful theory of strings has captivated many physicists and promised to explain all. Yet a major paradox in our understanding of gravity continues to taunt us.

We will explore gravity—its inner character and its outward manifestations. In the Prologue, we review Newtonian gravity. In Part 1, we trace our understanding of gravity from the old man's toy to the quantum age. From this discussion of gravity, we move naturally on to a glimpse of gravity at play and at work in the universe. In Part 2, we follow the universe from the big chill to the big bang. In Part 3, we consider how matter and structure emerged out of the void. In the evolution of the universe, the hand of gravity is sometimes direct and evident; at other times it is hidden behind the scenes but is just as indispensable. From this tour of the universe, we return in Part 4 to gaze at the innermost secrets of gravity. Thus, a book in four movements: gravity, universe, universe, and gravity.

This book is about gravity and about the universe. My ambition here is to explain the physics of gravity as well as the physics of phenomena involving gravity, as represented most dramatically by the dynamic universe. Cosmology is certainly of deep interest and shares equal billing with the physics of gravity. Cosmology is also considerably more accessible to the lay reader. Ultimately, however, the passion of the fundamental physicist lies more with the mystery of why we fall than with the life of the universe.

I would also like to say a few words in tribute to George Gamow, the late Russian-American physicist who fashioned modern physical cosmology out of the general notion of the big bang, as I describe in chapter 6. An ebullient and irreverent jokester, Gamow was notorious for the fun he managed to have while doing physics. Quite a few physicists now feel that the Nobel Prize committee passed him over unjustly. Perhaps they didn't like the way he treated physics as an amusing hobby rather than as a serious profession. Oh well, in any case, his style appeals to me. Gamow also wrote a delightful series of popular physics books. I went into physics partly because of a chance encounter with one of his books during high school. I recommend those books.

I did not want to write the kind of popular physics book that merely serves up headlines. As a physicist and a professor, I want to explain as much as possible. But in discussing quantum gravity and the superstring theory in Part 4, when the full subtleties of the quantum field theory come into play, I can, alas, do no more than give you the flavor of the physics involved. For those who want more, I can only suggest that you embrace a career in theoretical physics, as I did when I read George Gamow's confession in one of his popular physics books that he just couldn't explain quantum statistics.

On the other hand, I also want to report on the excitement of the latest developments. The trick is to avoid the kind of developments that are here today and gone tomorrow. Over the four years or so from the first writing to the publication of this book, the popular press has reported breathlessly on quite a few exciting "discoveries." By and large, I discuss in this book only what I believe will endure, at least in broad outline. When I do talk about speculative suggestions, I try to make clear that they are just that.

I have never liked how popular science books distort history and perpetuate the myth that a handful of individuals is responsible for all discoveries. Regretfully, I can't avoid doing the same: The name Einstein appears on virtually every page. I would have loved to recount the historical currents and influences leading up to Newton's great discovery, for instance, but I can do no more than hint at them. What little I can introduce can be regarded only as a sketch, if not a caricature, of history.

Acknowledgments

Bruce Caron, Pat Coakley, and Abd-al Hayy Moore, friends without any scientific background, read the manuscript. Their comments and questions helped enormously in making the book clearer. I am grateful to them.

My wife, Gretchen, also helped to eliminate obscurities. Her comments were crucial in clarifying certain difficult passages.

I am pleased to have David Wolff as my editor, and I thank him for his enthusiasm, encouragement, and patience throughout the project. His advice was indispensable in structuring this book.

Charles Levine, the editor for my previous book, *Fearful Symmetry,* was involved in this project in early 1985. I appreciate his support and editorial comments when I first started.

I am grateful to Jane Littell for her indispensable help in the final production stage of the book.

Finally, the questions of my copy editor, Linda Marshall, helped me to identify potentially confusing passages.

I thank Sonia Ng for arranging for some of the figures to be drawn.

FIGURE CREDITS

Stella Zee: Figures 6.2, 8.1, 12.1, 12.2
A. Zee: Figures P.6, 1.3, 2.2 2.5, 4.3, 4.5, 5.4, 5.5, 6.3, 9.1, 9.3, 9.4, 9.6, 9.8, 9.9, 9.10, 10.1, 11.3, 13.1

Figures 1.5a and 1.5b: *Explorers on the Moon* by Hergé. Courtesy Casterman Société Anonyme, Brussels, Belgium.
Figure 4.1: "The Constellation of Andromeda" from *The Bush of the Fixed Stars* by Ali-Sûfi (A.D. 903–986), reproduced from *The Search for the Nebulae* by Kenneth Glyn Jones, © 1975 by Alpha Academic, Canada. Used by permission of the author.
Figure 6.5: Photograph of Ylem from *My World Line* by George Gamow. Copyright © 1970 by The Estate of George Gamow. All rights reserved. Reprinted by permission of Viking Penguin Inc. Courtesy Per Anderson, American Institute of Physics, New York.
Figure 8.3: *A Midnight Modern Conversation* from *Hogarth's Graphic Works* by Ronald Paulson (plate 134), Yale University Press, 1970. Courtesy Yale University Press.

Prologue: The Apple and the Moon

A falling apple tells us that the same laws govern heaven and earth.

> I began to think of gravity extending to ye orb of the Moon & . . . I deduced that the forces wch keep the Planets in their Orbs must [be] reciprocally as the squares of their distances from the centers about wch they revolve: & thereby compared the force requisite to keep the Moon in her Orb with the force of gravity at the surface of the earth, & found them answer pretty nearly. All this was in the two plague years of 1665–1666. For in those days I was in the prime of my age for invention and minded Mathematicks & Philosophy more than at any time since.
>
> —NEWTON in his memoirs

THE PLAGUE

In the summer of 1665, England was visited by the plague. Cambridge University was closed, and the students were sent home. Among them was Isaac Newton, then twenty-three years old. He returned to his family farm at Woolsthorpe and spent the next two years there thinking and studying in rural isolation. One day an apple fell on his head. That fateful encounter between two solid objects inspired Newton to invent his theory of gravity.

Who will ever know if the story is true? It sounds as if it were fabricated by some scriptwriter. Actually, it was first told by Newton himself in his old age. In any case, it's a pretty good story, tinged with the Judeo-Christian imagery of the apple as the fruit of knowledge.

WHO WAS SIR ISAAC NEWTON?

Physicists revere Newton as the greatest of their profession who ever lived, and it is no exaggeration to say that his work laid the foundation of physical science as we know it. But who was Sir Isaac Newton?

This great genius was the first person on his father's side able to write his name. His father's brother saw no need to educate his sons, and they died illiterate. His mother was barely able to write and not particularly enthusiastic about sending Isaac to school. Newton was educated only at the urging of his mother's brother; that maternal uncle was in fact the only educated person in the entire clan.

Newton's paternal ancestors had risen within a century from obscure peasanthood to land ownership, a rise suggesting that they were above average in intelligence. Newton's father, said to be "a wild, extravagant, and weak man," died six months after his marriage. Newton was born fatherless some months later, on Christmas morning 1642. He was premature and so tiny that no one expected him to live.

When the fatherless boy turned three, his mother remarried, to a sixty-three-year-old clergyman. Bombastic and rather obnoxious, the reverend refused to let Newton live with his mother. Psychoanalysts have not hesitated to seize upon this traumatic loss of mother as an explanation for Newton's later behavior. As befitting the religious upbringing of the time, Newton kept note of his sins, and one of them was threatening to burn his stepfather and mother and their house.

When Newton was ten, his stepfather died. He was reunited with his mother but apparently felt intense jealousy, as his stepfather had in the meantime produced three children with Newton's mother. The reunion was brief as he was soon sent off to school in town. He lodged with an apothecary and apparently grew quite attached to the apothecary's stepdaughter. It was to be the first and last romantic attachment to a woman in his life.

When Newton turned seventeen, his mother decided that it was time for him to come home and help on the farm. Told to herd sheep, Newton would simply wander off and read. Court records showed that on several occasions he was fined for allowing his sheep to stray into the neighbor's fields. Newton's own list of sins for this period included "Punching my sister," "Peevishness with my mother," "Falling out with the servants," and "Striking many." The servants hated him and described him as a surly and lazy good-for-nothing.

Meanwhile, Newton's maternal uncle and the schoolmaster lobbied with his mother to let Newton go back to school. She relented only when the schoolmaster offered to waive the fee and to take Newton in at his own home. Newton's mother appeared to be quite a miser, as she was actually rather wealthy from the combined estates of her two husbands. When Newton went to Cambridge, he had to enroll as a subsizar, a poverty student who earned his way by performing menial tasks for the wealthier students. He fetched beer and emptied chamber pots, among other duties. The experience was mortifying for someone who grew up being surly to servants.

Richard S. Westfall, a leading Newton biographer, describes Newton as "a tortured man, an extremely neurotic personality who teetered always, at least through middle age, on the verge of breakdown." Small wonder, after that kind of boyhood. It doesn't take much psychoanalysis to see Newton's numerous bitter and nasty disputes with his scientific contemporaries as an expression of his infantile rage at his stepfather.

Newton was particularly ruthless in crushing his major scientific rival, Robert Hooke, seven years his senior and considered England's greatest scientist until Newton appeared on the scene. It is interesting to note in this connection that the often quoted line "If I have seen farther it is by standing upon the shoulders of Giants" is usually misinterpreted as evidence of Newton's modesty. In fact, the line was a sarcastic dig at Hooke, who was rather short and crooked in his posture. Some present-day physicists have been known to use the variant "If I have seen farther it is by looking over the shoulders of midgets" to put down their colleagues.

On the whole, Newton's years at Cambridge were the happiest of his life up till the plague years. Compared to the centers of learning on the Continent, Cambridge was an intellectual backwater at the time. The university had a prescribed curriculum based in part on Aristotelian studies, but fortunately by the 1660s the strict Aristotelian curriculum was not taken seriously, even at Cambridge. Newton, fortunately, was by and large left alone. He had time to reflect and to read what he wanted. By the time the plague came, he had already discovered the writings of Galileo and Descartes and invented the beginnings of calculus.

ANY TWO OBJECTS

By thinking on it continually.
—NEWTON (reply given when asked
how he discovered the law of gravity)

Newton discovered that any two objects attract each other. The more massive the objects, the stronger is this universal force of gravitation. As the two objects move farther apart, the attraction between them weakens, but it never quite decreases to zero.

As is well known by now, the apple falls because the earth is pulling it down. In fact, since not only apples but also all objects fall, the force apparently does not depend on the precise nature of the object, only on its mass—that is, the amount of "substance" contained in the object. How does this force depend on the mass?

GALILEO'S CANNONBALLS

At that time, Newton had already figured out the laws of motion. One of these laws says that when a force of any kind is exerted on an object, it causes the object to accelerate. Furthermore, the mass of the object times the acceleration is equal to the force exerted. $F = ma$: Force equals mass times acceleration. In other words, the acceleration of an object subject to a force is just the force divided by the mass of the object: $a = F/m$.

I should emphasize that Newton's law of motion $F = ma$ is about the effect of forces in general. The force may be electric, magnetic, gravitational, or whatever. Until Newton, people were blinded by ever-present friction into believing that a force is necessary to keep an object moving at a constant velocity. Perhaps by watching a skater gliding gracefully across a pond of ice, Newton realized that if friction were minimal, an object would maintain its velocity for a long time even without any force being exerted on the object. His law states that in the absence of friction a force exerted on an object changes the object's velocity—that is, it produces an acceleration.

Indeed, the gravitational force causes objects to accelerate as they fall. Already, Galileo had determined that falling objects accelerate at about thirty-two feet per second per second. That means that after falling for a second, a falling object is moving at a speed of thirty-two feet per second; after two seconds, sixty-four feet per second; and so on. Of course, air resistance would slow it down some.

Given all this, Newton deduced that the earth's gravitational pull on an object is proportional to its mass m; in other words, F is equal to m times some constant. Call the constant g for gravity. Thus, $F = mg$. The proof is to show that this accords with Galileo's observations.

That's easy. The acceleration is $a = F/m = mg/m = g$. The mass m cancels out! The acceleration does not depend on the mass of the object!

But that's exactly right! Galileo said that the acceleration of a falling object is thirty-two feet per second per second. Period. He didn't have to specify what the object's mass is.

To underline this key point, let us suppose erroneously that the gravitational force is proportional to the mass squared, for example. Then the acceleration of a falling object would be $a = F/m = $ constant $xm^2/m = $ constant xm. Thus an object twice as massive as another would accelerate twice as much as the other. That would contradict Galileo's observations.

Supposedly, Galileo demonstrated that all objects fall at the same rate by climbing up on the Leaning Tower of Pisa and dropping off two cannonballs of different sizes, showing that they hit the ground at the same time. Most people refused to believe Galileo, asking him how a rock and a feather could be said to fall at the same rate. These people didn't appreciate that air resistance on the feather far exceeds that on the rock. It is a dramatic and true fact that in a vacuum a rock and a feather would fall side by side.

As we will see, that falling objects fall at the same rate provides the clue to the true nature of gravity and the cornerstone on which Einstein will construct his theory of gravity. From this fact flow the profound secrets of space and time.

PULLING THE EARTH UP

Strictly speaking, it is not correct to say that the earth is pulling the apple down. In their gravitational *pas de deux*, the earth and the apple attract each other. As the earth pulls the apple down, the apple also is pulling the earth up. When I first heard of this as a kid, this made quite an impression on me. Every time I fall out of a tree, the earth is actually falling toward me. The same force that acts on the apple also acts on the earth. Of course, with the same force, the earth moves much less than the apple, since the earth is so much more massive.

By the same token, as we walk we are also pushing the earth backward. Think of your heels pushing the landscape backward just as when you swim you push the water backward. At any given time, of course, the effects of all the people and vehicles moving about on the earth, minuscule to begin with, pretty much cancel out.

THE LONG FINGERS OF GRAVITY

We just decided that the gravitational force between the earth and an object of mass m is proportional to m. Is that gravitational force also proportional to the earth's mass? It turns out that it is. This fact, though vaguely plausible, is not obvious, however, because the earth is so much more massive than an ordinary falling object.

The issue of whether the gravitational attraction between the earth and the apple is also proportional to the earth's mass is closely related to the other question Newton had to answer, that of how the attraction between two objects depends on the separation between them.

Consider the possibilities. The force between two objects could decrease abruptly to zero as we separate the two objects beyond a certain distance, called the range or the reach of the force. Beyond this range, the force would simply cease to operate. Suppose for the sake of argument that gravity has a range of a hundred yards. Then only that portion of the earth within a hundred yards of the apple would contribute, and the gravitational force between the apple and the earth would certainly not be proportional to the entire's earth's mass. But if gravity has a range much larger than the size of the earth, then the entire earth's mass could well contribute. An extreme possibility is that as the distance between two objects increases, the gravitational force between them decreases gradually toward zero but never quite becomes zero. In that case, gravity would have an infinitely long range.

To determine gravity's range, Newton proceeded with the simplest assumption that the gravitational attraction between two objects of masses m_1 and m_2 is indeed proportional to both m_1 and m_2. The force doesn't favor one over the other. We will have to come back and verify that that is in fact true, even between the apple and the earth.

THE ROCK IN THE SKY

How could Newton possibly determine how the gravitational force depends on the distance between the two objects? In principle, he could try to drop the apple from greater and greater heights and see how the gravitational force exerted on the apple decreases. The trouble is, the earth's radius is about four thousand miles. Any height Newton could possibly reach, when viewed from the perspective of the entire earth, was going to be tiny. There was no way in which he could take the apple to a height of thousands of miles from the ground. Of course, it doesn't have to be an apple. Any massive object would do—a rock, for example. But he couldn't get a rock up there, either.

But there is already a rock up there. The moon! This was Newton's brilliant insight.

For an object, such as the moon, to orbit another object, such as the earth, a force must be exerted on the moon, pulling it inward. If not, the moon would fly off in a straight line. In Figures P.1a and 1b, a boy

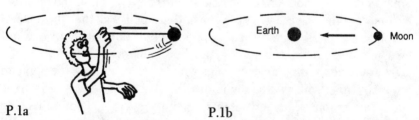

P.1a   P.1b

P.1a. *A boy whirls a stone tied to a string. To keep the stone from flying off, the boy must exert a force on the stone by pulling on the string. The faster the stone whirls, the stronger is the force needed.* **P.1b.** *For the moon to follow its orbit around the earth, the earth has to exert a force on the moon. The same physical principles are involved in* **P.1a** *and* **P.1b**.

ties a stone to a string and whirls it around his head as a slingshot. To keep the stone going round and round, he has to exert a force on the stone by pulling on the string. The moment he lets go, the force disappears and the stone goes flying off.

Newton must have realized with a start that the gravitational force exerted on the moon by the earth was just what is needed to keep the moon on its appointed rounds. Using his laws of motion, Newton could figure out how large this force had to be: It depends on the radius of the moon's orbit and the speed with which the moon moves around in its orbit.

The same discussion applies to the motion of the planets around the sun. In fact, the astronomer Kepler had already painstakingly determined the radius of each planet's orbit and the speed with which each planet moved around in its orbit. And so Newton knew the force needed to keep each planet in orbit and thus the force exerted by the sun on each planet.

What a stroke of luck! Nature provided Newton with a whole bunch of planets at different distances from the sun. He could finally determine how the gravitational force between two objects depends on the distance between them. Using Kepler's data, Newton found that the gravitational force between two objects decreases as the square of the distance between them. If we double the distance between the two objects, the force decreases to $1/(2 \times 2) = \frac{1}{4}$ of what it was. If we triple the distance, the force decreases to $1/(3 \times 3) = \frac{1}{9}$ of what it was. And so on. The gravitational force is infinitely long-ranged.

That gravity is long-ranged (physicists usually drop the adverb "infinitely") has enormous implications for the evolution of the universe. An object in the universe feels the gravitational pull of all the other objects in the universe. Of course, each far-off object would contribute a minuscule amount, but the total effect may be significant.

Gravity's long range explains why the force causing the apple to fall is so strong while we hardly notice the attraction between the apple and a single stone. The gravitational force is noticeable because its long range allows the entire mass of the earth to contribute.

Trumpets, fanfare, please! Newton could now put it all together into a law of universal gravitation. The gravitational force between any two objects, of mass m_1 and m_2, respectively, and separated by the distance d, is equal to Gm_1m_2/d^2. The force is equal to the product of the two masses and a constant, *G*, now known as Newton's constant, all divided by the distance squared.

The constant *G* measures the overall strength of the gravitational force. With all the data available to him, such as the known size of the earth, Newton could determine the value of *G*. In fact, this then gave him a way of estimating the sun's mass.

Whew! All this from the falling apple.

TO THE CENTER

After Newton formulated his theory of gravity, he recognized that he had to go back to the force between the apple and the earth. In his formula, should he put in the distance from the apple to the surface of the earth, or to the center of the earth, or to some other point inside the earth? Also, should he put in the entire mass of the earth, or does only a portion of the earth's mass contribute?

Fortunately, these questions did not come up when Newton used the motion of the planets around the sun to find the inverse square distance dependence. On the scale of the solar system, the separation between the sun and any planet is so large compared to the sizes of the sun and the planet that the sun and the planet may be viewed as tiny balls and idealized as mathematical points. The relevant distance is clearly the distance between the points. On the scale of the falling apple, however, the earth is so large that it can hardly be treated as a point.

Newton knew what to do. Imagine cutting the earth into tiny pieces, so tiny that each piece can be treated almost as a mathematical point. Calculate the gravitational pull exerted on the apple by each of these pieces, using Newton's formula. Add the gravitational pull due to all these pieces together. There you have it: the gravitational pull exerted by the earth on the apple.

The result is simple to state. To calculate the gravitational force between the apple and the earth, we use Newton's formula and put in the distance from the apple to the center of the earth. We also put in the entire mass of the earth.

We can understand roughly why. Suppose the apple is at the North Pole. Imagine the earth as a spherical cake. Slice it into pieces, cutting in at the latitude lines, at ten-degree intervals, moving the knife perpendicular to the north–south axis. (See Figure P.2.) The apple is pulled down by all these pieces, first by a tiny cap containing the frozen

P.2. *While the Arctic cap is closest to the apple, the equatorial pieces are much more massive. Effectively, the earth pulls on the apple as if the earth's entire mass is concentrated at its center.*

Arctic, then by a slightly bigger piece containing the northern tips of Siberia and Greenland, and so on. The pieces get larger until we reach the equator and then get smaller again as we head toward the South Pole. Each of these pieces pulls on the apple. Although the Arctic piece is closer to the apple, the two equatorial pieces contain a lot more mass. Thus they have the most pull. Hence it is plausible that when the pulls of all these pieces are added together, the earth's gravitational pull turns out to act effectively from its center. (Obviously, there is nothing special about the North Pole. In the discussion, I chose the North Pole only because the geography is easier to describe.)

For this simple result to hold, it is crucial that the gravitational force decreases as the square of the distance. The decrease is just right so that the entire earth contributes: The mass of the entire earth is to be used in Newton's formula. Suppose, on the contrary, that the gravitational force decreases much more drastically than the square of the distance. Then the Arctic piece would have the most pull. Even though the equatorial pieces are larger and more massive, their effects are cut down by their distance from the apple. In that case, not all of the earth's mass contributes, and the relevant distance would not be the distance from the apple to the center of the earth.

Incidentally, Newton did not publish his theory of gravity for some twenty years after he first conceived it. One reason is that apparently he did not know right off how to prove this result we just stated, even though he suspected it to be true. The notion of dividing an object—the earth, in this case—into many tiny pieces and then adding up the effect due to each piece is one of the central ideas of calculus. In 1666, Newton hadn't yet invented this portion of the calculus.

NOT PULLED HITHER AND YON

When I go near a massive object such as a mountain, shouldn't I feel the gravitational pull of the mountain? When I first heard about Newton's law of gravitation, I naturally wondered about this question. Sure, the mountain is tiny compared to the earth, but the center of the earth is awfully far away compared to the center of the mountain. Is mass or proximity more important? We now know enough to figure this out.

For a rough estimate, the precise shape of the mountain shouldn't matter. For simplicity, take the mountain to be a ball with a radius of 4 miles. Suppose you are standing right on the mountain. (Figure P.3.) Since the radius of the earth is 4,000 miles, and since the gravitational force decreases as the square of the distance, the earth has to concede in proximity a factor of $1,000 \times 1,000$ to the mountain. Can the earth make

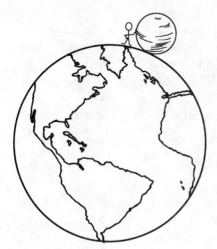

P.3. *A spherical mountain on earth. The gravitational pull of the earth is at least a thousand times stronger than the gravitational pull of the mountain. (The figure is not drawn to scale, of course.)*

it up by being more massive? Yes, it can. Since its radius is 1,000 times larger than the radius of the mountain, the volume of the earth is 1,000 × 1,000 × 1,000 times larger than the volume of the mountain. Again, roughly, the earth and the mountain are made of the same rocks and stones and thus have more or less the same density. Thus, in massiveness, the mountain has to concede a factor of 1,000 × 1,000 × 1,000 to the earth. Putting it together, you see that your gravitational attraction to the earth is larger than your gravitational attraction to the mountain by a factor of 1,000 × 1,000 × 1,000 divided by 1,000 × 1,000—that is, by a factor of 1,000. Mass wins over proximity. Fortunately so. Otherwise it would be quite bothersome as we would be pulled hither and yon by every massive object we pass by. From this estimate, you can also see that the gravitational attraction two persons exert on each other is absolutely tiny.

THE GRAVITY EXPRESS

When I was in high school, I read about how gravity might be exploited to create a new mode of travel. Suppose a tunnel could be dug straight down through the center of the earth to the other side. Look down into it. Scary! An apparently bottomless pit. We have bought our ticket and now we are invited in. We take a deep breath and we jump.

Whoosh! We fall straight down, just as Alice once fell into the rabbit hole. We go faster and faster. It is that falling dream come true! We go rushing past the center of the earth. The moment we go past the center, gravity starts to act as a brake. It tries to pull us back to the center. We start to slow down. Will our momentum carry us through? Yes! Provided there is no friction, we would just make it to the exit hole on the other side. We would also need help getting out of the hole, lest we fall right back in.

In practice, there is friction, and we would need a little boost to get us to the end. Of course, we would also need to be inside a capsule strong enough to withstand the searing heat at the center of the earth, not to mention the lightning speed we will get up to as we zoom past the center.

Since we know the strength of the gravitational force, we can easily figure out how long the trip would take. It turns out to take forty-two minutes! No airplane with current technology could possibly get you there that fast. Better yet, no fuel is needed for this "gravity express." We just let gravity do its thing.

You are suspicious. What is the catch? There is no catch, aside from the expense and engineering difficulty of building a heatproof tunnel that would not collapse through the center of the earth. The physics is perfectly sensible. Gravity always is trying to pull us down; we will just let it!

GRAVITY INSIDE THE EARTH

To understand in a little more detail how the gravity express works, we need a result first obtained by Newton. He posed the following apparently irrelevant question. Suppose we find ourselves somehow inside a large spherical shell. (Figure P.4a.) The shell, being massive, exerts a

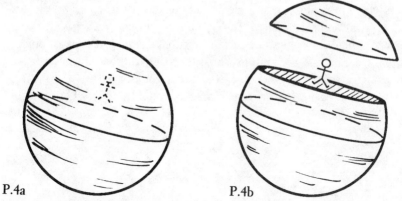

P.4a   P.4b

P.4a. *How large is the gravitational force exerted by a spherical shell on a person? (While the illustrator wants to show the location of the person, he is also eager to indicate that we cannot see through the solid spherical shell. He uses a dotted line as a compromise.) In **P.4b**, to indicate that we are talking about a spherical shell (and not a solid ball), the illustrator imagines cutting the shell into two caps. The caps are then pulled apart to show that the sphere in **P.4a** is hollow and to reveal a person. The gravitational force exerted by the two caps on the person balances each other out.*

gravitational force on us. In which direction does this gravitational force pull us?

Imagine someone cutting the spherical shell in two. Clearly, the rims of the two pieces form a circle. The cut is to be done at such an angle that we are at the center of the circle. One picture is worth a thousand words; the division is shown in Figure P.4b. For clarity and to reveal our position, the illustrator has pulled the two pieces apart slightly.

Let us simply call the two pieces the small piece and the large piece. We have to compare the gravitational pull exerted on us by the small piece with the pull exerted on us by the large piece. It is a tug-of-war. The small piece contains less mass, but on average, the matter in the small piece is closer to us than the matter in the large piece. It is plausible that the pulls exerted on us by the two pieces balance out. Newton did the calculation and found that that was indeed true. The spherical shell does not exert any net gravitational force on us.

This is a nice little result, but how is it relevant to a discussion of the gravity express? We would like to know the gravitational pull on us when we are inside the earth at the point A indicated in Figure P.5a. Well, mathematically we can think of the earth—or any ball, for that matter—as formed of many spherical shells nested one inside another, sort of like how an onion is formed. In Figure P.5b we show a cutaway view. Since a spherical shell does not exert any gravitational force on an object inside it, we can, for the purpose of calculating the gravitational force exerted on us, simply get rid of all the outer shells.

This leads us to the remarkable conclusion that when we are

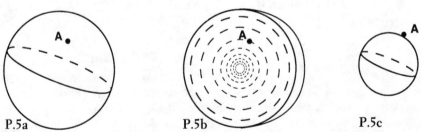

P.5a   P.5b   P.5c

P.5a. *What is the gravitational pull on us when we are inside the earth at the point A?* **P.5b.** *We can think of the earth, as an onion: formed of many spherical shells. Here a cutaway view is shown.* **P.5c.** *When we are at the point A, we only feel the gravitational pull of the inner core of the earth, as explained in the text.*

inside the earth, only the inner core exerts a gravitational pull on us. The inner core is defined as consisting of all the matter that is closer to the center of the earth than we are. Thus, when we are on the earth's surface, as we normally are, the inner core is the entire earth. As we descend toward the center of the earth, the inner core forms an ever smaller ball. At the center of the earth, the inner core shrinks to nothing. Now we know how to calculate the gravitational force we will feel; it is G times our mass times the mass contained in the inner core divided by the square of the distance from where we are to the center of the earth. This is illustrated in Figure P.5c.

This result makes sense. As we descend toward the center of the earth, the inner core gets smaller and smaller and thus the gravitational pull on us decreases gradually from its normal value at the surface of the earth to zero at the center. This is just what we might expect, since it is obvious that when we arrive at the center of the earth we should feel no gravity at all. We are pulled outward equally in all directions, and the effects cancel out.

Incidentally, the gravity express can connect any two cities. We dig the shortest tunnel joining the two cities. (Figure P.6.) Now the

P.6. *The underground gravity express: an efficient mode of transportation if it could be built.*

tunnel does not go straight down. We slide down at an angle rather than fall straight down. Thus we go slower. On the other hand, there is less distance to cover. It turns out that the trip takes forty-two minutes regardless of which two cities we connect.

Although we are not going to journey to the center of the earth anytime soon, what we have learned here will be of central importance when we discuss the universe and galaxies. To see how, consider a more or less spherical galaxy. (The number of stars in a galaxy is so large that on the scale of the galaxy, matter can be thought of as spread out throughout space.) The problem of determining the gravitational force felt by a star is mathematically identical to that of determining the gravitational force we felt on our journey to the center of the earth. Take a star—say the one indicated by A in Figure P.7.Imagine drawing a sphere about the center of the galaxy and with a radius such that the star sits on the surface of the sphere. According to what we have just learned, only to the matter enclosed within that sphere pulls on the star gravitationally. The gravitational effect of the matter outside that sphere adds up to zero.

By observing the movement of stars and hence deducing the gravitational force they feel, we can map out how matter is distributed inside the galaxy. I will say much more about this later.

CULTURAL CONTEXT

The actual path by which Newton arrived at the law of gravity is, of course, considerably more arduous than what I have indicated. Since

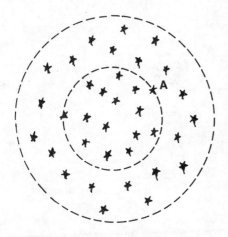

P.7. *In a roughly spherical galaxy, only the matter inside the sphere indicated here by a circle pulls gravitationally on the star A. Compare with Figure P.5. (The figure is not drawn to scale of course. A typical galaxy contains many more stars than shown here. Also, this is a two-dimensional drawing of a three-dimensional situation.)*

Newton's notebooks are well preserved, generations of Newtonian scholars have analyzed Newton's thought process in the cultural context of his time.

At that time there was still a tremendous amount of confusion over the concept of force. Many thought that force was internal to an object, some sort of inner vigor. Newton was also unsure whether action at a distance was possible. Many natural philosophers of the time—Descartes, for instance—thought that one object can exert a force on another only if the two are in contact. At one point, Newton thought that gravity was the result of a steady stream of tiny, invisible particles raining down on the earth. Objects fell because they were driven downward by this steady rain. (This explanation, of course, begs the question of what is pulling the tiny particles down. It would also indicate that larger objects would fall faster.) Even after arriving at the correct law, Newton used a wrong value for the size of the earth and thus obtained too small a value for the earth's gravitational pull on the moon. For some time, he thought that some other force besides gravity was necessary. It wasn't easy.

To appreciate Newton's genius, we must recall that in his day people thought of the heavenly bodies as moving in another realm and spoke rhapsodically of the music of the celestial spheres. While the motions of earthly objects are always rudely brought to a stop by the relentless drag of friction, the celestial bodies cruise on unceasingly for all eternity. Viewed in this context, it can be rightly said that Newton's leap of genius lies in extending gravity, an apparently terrestrial phenomenon, all the way out to "ye orb of the Moon." After Newton, the same laws of physics govern heaven and earth. In this sense, Newton's law of gravity may be regarded as one of the cornerstones of modern civilization.

THE BEGINNING OF PHYSICS

Physics can be said to have begun with Newton's theory of gravity. When I reflect on Newton's understanding of gravity, I am struck by how all the circumstances, scientific and natural, were propitious. Galileo and Kepler had compiled precisely the data that were needed. Newton himself had just mastered the laws of motion. The natural circumstances were right. The earth has a moon. Perhaps Newton had sat till nightfall under the apple tree thinking about why the apple fell. Looking up, he might have seen the moon and immediately wondered why it didn't fall. The moon took him from earth to heaven, from terrestrial physics to celestial physics. From the moon, his mind voyaged out to the planets. And there was more than one planet, at various distances from the sun, and thus he could practically read off the distance dependence of the gravitational force.

Isaac Asimov once wrote a memorable story describing a civilization in a multiple-star system. Somewhere, several stars happen to be clustered together, and they orbit around each other, following some complicated paths. In this melee weaves a lonely planet without a moon. Standing anywhere on this planet, you always face at least one star. As one sun sets, some other sun has already risen. The civilization that grows up on this planet does not know what night is. The civilization evolves to technological sophistication, but it does not understand why apples fall. Because of the complicated orbital dynamics, it has no inkling of the fundamental force of gravity. Finally, a young genius realizes that by postulating gravity, the complicated schedule of all those suns rising and setting can be immediately understood. He can track the movement of all those suns, and to his surprise he discovers that every ten thousand years the suns will line up on one side of the planet. The sky will go dark. We can imagine an ending for the story. His warnings are ignored. Finally night falls. The sky goes black. And civilization disintegrates in a firestorm of mass hysteria.

Physics began with gravity, but it may also end with gravity. Newton's theory merely describes how gravity operates. As we will see, the fundamental mystery of gravity continues to haunt physics.

THE RISE OF GRAVITY

I

We trace our understanding of gravity from the old man's toy to the age of the quantum.

An Old Man's Toy

HIS LAST BIRTHDAY

On Einstein's seventy-sixth and last birthday, March 14, 1955, his neighbor Eric Rogers presented him with a toy constructed of a heavy ball, a spring, a broomstick, and other commonly found objects. The contraption is shown in Figures 1.1a, 1b, 1c. A brass ball attached to a string hangs outside a metal cup into which the ball can fit snugly. The string passes through a hole in the cup and down through a pipe, where it is tied to a spring. This entire assembly is mounted on a curtain rod so that one can hold on to the whole contraption easily. Finally, the cup and ball assembly is enclosed in a transparent glass sphere to give it a finished look.

    Were the spring strong enough, it could pull the ball into the cup. By design, however, the spring is too weak to counteract the force of gravity. And so the ball hangs limply outside the cup.

    By shaking the curtain rod, it's possible to pop the ball into the cup. But with a flabby spring and a small enough cup, this may be made frustratingly difficult. The challenge is to find a surefire way to pop the ball into the cup every time.

    Einstein was delighted. He recognized immediately that the necessary trick hinged on a physical principle he himself had thought up half a century earlier. He was pleased that his friend Professor Rogers took the trouble to remind him in this fun way of what he had described as one of the happiest moments of his life, the moment when he understood gravity. In his delight, the old man's mind momentarily flashed back.

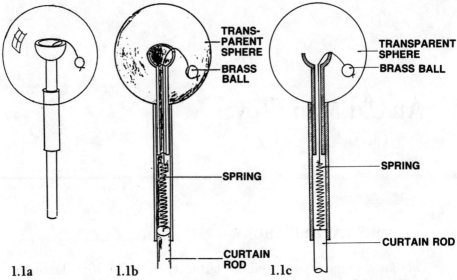

1.1a   1.1b   1.1c

The old man's toy. For maximum clarity, the artist shows three different views of the toy. 1.1a. *A three-dimensional rendering.* 1.1b. *A three-dimensional cutaway drawing.* 1.1c. *A two-dimensional cross section.*

A HAPPY THOUGHT

Happiness goes like the wind, but what is interesting stays.
—GEORGIA O'KEEFFE

The scene of the birthday gathering dissolves, and we find ourselves in the summer of 1900. Albert Einstein, together with three other students, had just finished their physics studies at the technical university ETH (Eidgenössische Technische Hochschule) in Zurich. While the authorities immediately appointed the other three students as assistants, they regarded the dreamy Einstein as unsuitable for a university position. Dejected, Einstein wrote letters all over, trying to get a job, but without success. He wrote later that he felt like "a pariah, discounted and little loved . . . suddenly abandoned by everyone, standing at a loss on the threshold of life."

Einstein spent the next two years in frustration, holding temporary jobs as a tutor and as a high-school substitute teacher in various small towns in Switzerland. Finally, in early 1902, he heard of an opening for a technical expert second class at the federal patent office in Bern, and as luck would have it, Einstein had a close friend whose father knew the director of the patent office. With a certain amount of string-pulling, Einstein was called for an interview. After a two-hour grilling, the official examiner noted for the record that the examination "unfortunately disclosed [Einstein's] obvious lack of technical training." Nevertheless, the much-maligned old boys' network triumphed, and in June 1902 Einstein was hired, but only as a technical expert third class.

Einstein was overjoyed. Now finally employed, he was able to marry Mileva Maric, a school friend from his Zurich days, and to start a family. He liked his job. The work was interesting and also left him time to think about his beloved physics. Bern was a university town, and Einstein soon befriended several young men with similar interests with whom he met regularly for intellectual discussions.

In 1905, Einstein shook physics to its foundation with his papers on the special theory of relativity. A year later, he was promoted to technical expert second class. Then one day in November 1907, inspiration struck: He had what he later called the happiest thought of his life.

"I was sitting in a chair in the patent office at Bern when all of a sudden a thought occurred to me: 'If a person falls freely he will not feel his own weight.' I was startled. This simple thought made a deep impression on me. It impelled me toward a theory of gravitation."

From this apparently nonsensical idea that a falling person feels no gravity emerged the secrets of gravity and the universe.

AN APRIL FOOLS' PRANK

The purest mobile form, the cosmic one . . . is only created through the liquidation of gravity.

—PAUL KLEE

5

What was the patent clerk thinking about? Let us try to understand.

A sky diver jumps out of a plane and feels an exhilarating floating sensation. Her mind tells her that eventually she will reach solid ground; her parachute had better open or she will be turned to mud. But for now, she can close her eyes and enjoy the floating sensation. She knows that she is falling because of the air rushing up past her. Imagine removing the air. Were it feasible for a sky diver to fall through a pure vacuum, she would not know she is falling. Hearing nothing but the pulse of her blood, she would not feel the force of gravity at all.

To remove the effect of air rushing up past the sky diver, put her inside a box. In fact, let us go all out and play an elaborate April Fools' Day prank on one of our friends. Drug the poor victim in her sleep and put her in a spacious box elaborately furnished inside to look exactly like her living room. We then drop the box from a high-flying airplane. (See Figure 1.2)

When our friend wakes up, she will think that she is in her living room. Curiously, though, she feels that she is floating. To an observer on the ground (assume that the box is such that one can see in but not see out), our friend and her living room are hurtling toward a crunching rendezvous with the ground. Our friend, however, is blissfully unaware of the impending disaster. Since she is accelerating downward at the same rate as the box and all the objects contained inside, she feels that she is not moving downward at all relative to her surroundings. A slight spring in her step and she finds herself drifting toward the ceiling. She feels that she is floating—but this action is interpreted by the ground observer quite differently: Our friend, by stepping on the floor, has at the same time decreased slightly her downward velocity and increased slightly the box's downward velocity. Our friend thinks she is floating upward but in reality her downward plunge is accelerating at the same rate as before.

Skeptical, you say that all this sounds like a physicist or a patent clerk playing with words. A falling person doesn't know she is falling?! Tell me a better one. Surely, if I fell out of an airplane, even if I were inside a box, I would *feel* the force of gravity dragging me down!

I can easily overcome your skepticism: The awfully unethical April Fools' joke has already been tried. Surely, in the comfort and security of your own home, you must have seen on television the antics of astronauts floating in their spaceships orbiting the earth.

1.2. *An April Fools' prank.*

But, you object, the space shuttle is safely orbiting the earth, not falling freely.

Well, get ready, I am going to hit you with yet another apparently paradoxical statement: The orbiting shuttle is staying up there precisely because it is freely falling. Let me explain.

STAYING UP BY FALLING

An enormous but common misconception is involved here. By now, everyone knows that the astronauts float because there is no gravity in space. Right?

Wrong! Surely, you don't believe everything television announcers say. Since the gravitational attraction between two bodies decreases as the distance between them increases, then deep in space, far away from any massive object, there is indeed no gravity. But the shuttle is orbiting the earth a mere one hundred miles or so aboveground. Using the scales of everyday life, we naturally think of one hundred miles up as very high up indeed. After all, Mount Everest is only about five miles high. However, in this context one hundred miles is an insignificant height. The gravitational pull on the shuttle and on everything inside it is just about as strong as it would be were the shuttle parked on the ground.

The point is that in calculating the gravitational force on the shuttle, what counts is not the distance from the shuttle to the surface of the earth, but the distance from the shuttle to the center of the earth, as we learned in the Prologue.

The radius of the earth is about four thousand miles. Thus, as the shuttle goes from its ground position to its orbital position one hundred miles above our heads, the distance from the shuttle to the center of earth increases from four thousand miles to forty-one hundred miles, a mere 2.5 percent increase. Recall that the gravitational force between two bodies decreases inversely as the square of the distance between them. As the shuttle goes into orbit, the earth's gravitational pull on it decreases by a factor of $(4000/4100)^2$, about a 5 percent decrease. The gravitational pull on the shuttle and on the astronauts inside it is almost as strong as would have been the case had they just been sitting on the ground.

The astronauts are definitely falling (although at a rate only 95 percent of the rate they would experience falling out of a tree). But yet they do not feel they are falling, because the shuttle around them and everything else inside it is also falling. In the same way, our friend does not feel that she is falling because her furniture and everything else around her is also falling.

A CLEVER FELLOW

To see how it makes sense to say that the shuttle is constantly falling while it stays up there in orbit, we have to go back to Isaac Newton.

Newton, clever fellow he, figured out, using the following ingenious argument, how a satellite can be put into orbit. Imagine placing a cannon on the highest mountain peak you can find. Fire, and note the distance the cannonball travels before it strikes the ground. Increase the firepower and the cannonball will travel farther. (See Figure 1.3a) What would happen, Newton asked himself, if I keep on increasing the firepower? Obviously, the cannonball goes farther and farther.

But wait—Newton realizes there is a catch. The picture we just drew is not quite right. The earth is round! As the cannonball flies farther

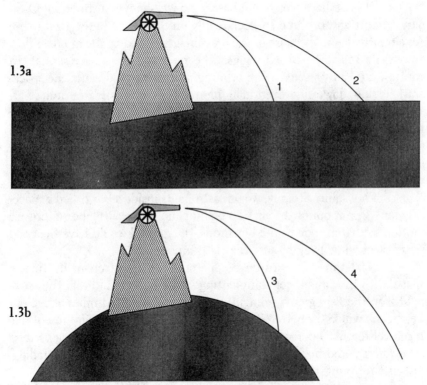

1.3a. *If the earth were flat, a cannonball fired from a mountaintop would eventually crash to earth (trajectory 1). As you boost the firepower, the cannonball travels farther (2) before it crashes to earth.* 1.3b. *On the round earth, however, the ground "tries" to get away from the cannonball (3) as the cannonball "tries" to fall to earth. A fast enough cannonball (4) would end up staying in orbit.*

and farther out, the curvature of the earth becomes more and more important. We have a curious situation: The cannonball is trying to fall to ground, but the ground is trying its darndest to get away from the cannonball. (See Figure 1.3b.)

Newton realized that if the ground is "dropping away" as fast as the cannonball is falling, the cannonball is in orbit.

Satellites and shuttles stay "up there" in orbit not because they are *not* falling. They are falling all right, day and night, seven days a week. Their trick is to move forward fast enough so that by the time they have fallen by a foot the ground has also dropped away by a foot.

(It is perhaps worth emphasizing a well-known distinction: Ordinary aircraft and orbiting spacecraft rely on different physics to stay up. An aircraft stays up because of the pressure generated by the rapid air flow around its specially shaped wings. The presence of air is essential. In contrast, any air present at the orbiting altitudes would cause the spacecraft to slow down and eventually plunge back into the atmosphere.)

When I first came across Newton's argument, I was impressed. It is one thing to talk about centrifugal force, vectors, kinetic energy, and so on, but to say that a satellite is staying up there because the ground is dropping away represents an extra measure of inspiration and understanding.

Thus, quite literally, we put astronauts inside a box called a spaceship and drop it out of the sky. To be humane, we give the box a forward motion so that as soon as the box drops, the ground would have the good sense of curving away by just the right amount.

Next time you turn on the TV and see an astronaut floating in space, with the announcer commenting in the background that the astronaut is in the zero-g environment of space one hundred miles above our heads, you will know better. The astronaut is in a 0.95-g environment. He is subject to only 5 percent less gravity than we are. *He is floating because he is falling,* and because he is falling he does not feel that he is falling, just as the young technical expert second class thought.

A BIG GRIN

We can now see how the old man's toy illustrates this fundamental truth about gravity. The eminent historian of science I. Bernard

Cohen visited Einstein not long after his birthday and wrote about his visit:

> At last I was taking my leave. Suddenly [Einstein] turned and called "Wait. Wait. I must show you my birthday present."
>
> Back in the study I saw Einstein take from the corner of the room what looked like a curtain rod five feet tall, at the top of which was a plastic sphere about four inches in diameter. "You see," said Einstein, "this is designed as a model to illustrate the equivalence principle . . ."
>
> A big grin spread across his face and his eyes twinkled with delight as he said, "And now the equivalence principle." Grasping the gadget in the middle of the long brass curtain rod, he thrust it upwards until the sphere touched the ceiling. "Now I will let it drop," he said, "and according to the equivalence principle there will be no gravitational force. So the spring will now be strong enough to bring the little ball into the plastic tube." With that he suddenly let the gadget fall freely and vertically, guiding it with his hand, until the bottom reached the floor. The plastic sphere at the top was now at eye level. Sure enough, the ball rested in the tube.

(The idea that a falling object does not feel any gravity is part of the *equivalence principle*, about which more later.)

The little brass ball is just as easily fooled as an astronaut. When Einstein let his toy fall, the little brass ball, precisely because it was falling, did not feel any gravity; the ball was the stand-in for the falling person in the patent clerk's daydream. The spring, normally too weak to pull the ball up against gravity, now seized the chance to yank the ball into the bowl. Get it? It is really quite simple. As you are falling, you are not aware of gravity. Of course, as an animate being, you are concerned about how the eventual impact will hurt. But that is psychology, not physics.

THE FALLING CANDLE

Einstein loved to pop playful little puzzles on his visitors. He was equally delighted whether or not they knew the answers. If they didn't, he would get a big kick out of explaining it. Here is one that he asked on more than one occasion.

Suppose you have just lighted a candle in an elevator when, unfortunately, the cable breaks. The elevator falls freely. What happens to the candle flame?

First of all, we have to understand how a burning candle works normally. The hot gas produced by the burning candle, being less dense than air, rushes upward. The upward rush of the glowing gas is what we see as the flame. The candle is thus assured of a steady supply of oxygen from the ambient air as the gas rushes out of the way. The second point is that the upward rush of the gas can be better interpreted as due to gravity pulling the denser air down. By moving downward, the ambient air is actually displacing the gas upward.

Good. We can now answer the grinning old man looking at us with a twinkle in his eyes. The falling candle feels no gravity, and neither does the air around it. The hot gas expands outward rather than rushing upward out of the way. For a moment, the candle is deprived of air supply and goes out.

UNIVERSALITY

To understand gravity in more detail, let us consider our April Fools' prank again. For the prank to work, it is crucial that all objects fall at exactly the same rate. Suppose to the contrary that the box falls faster than our friend. Then our friend would find herself pinned to the ceiling, which she would interpret as being due to the presence of a force pushing her up. Conversely, if the box were to fall slower, our friend would feel a force pulling her to the floor. The extreme case in which the box is not falling at all is of course the normal situation, with the box resting on the house foundation.

That objects all fall at the same rate regardless of their composition is contrary to everyday intuition, but as Galileo suspected, our everyday experiences are distorted by air resistance. In a vacuum, a feather and a cannonball fall at the same rate. Hard to believe, but true. Remember how Galileo is supposed to have demonstrated this by dropping several objects simultaneously from the Leaning Tower of Pisa? (Incidentally, the temptation to drop things off the Leaning Tower is so overpowering—I was barely to able to restrain myself—that I could well believe that Galileo actually did it, even though historians now say the whole story is apocryphal.) In the three hundred–odd years since Galileo, physicists have performed ever more sophisticated versions of his experiment with ever increasing accuracy. That all objects fall at the same rate is extremely well verified.

Physicists refer to the fact that different objects all respond in exactly the same way to gravity as the universality of gravity.

The gravitational force acts on all objects equally; it is universal. In sharp contrast, the electromagnetic force is not universal. Indeed, the electric force does not act on objects with zero charge, while it pushes positively charged objects and negatively charged objects in opposite directions. Also, when acted upon by the electric force, an object with twice the electric charge as another (but with the same mass) will accelerate twice as much.

Any deep theory must be able to account for the universality of the gravitational force. In Newtonian physics, universality follows from the law of motion and the supposition that the gravitational force on any object is proportional to its mass, as we saw in the Prologue.

But let us go back and try to imagine the patent clerk's train of thought. A falling person does not know she is falling, because everything around her is falling at the same rate; in other words, because of universality. Hey, wait a minute, can I turn this around? Gravity must be universal because a falling person does not know she is falling. In a way, falling cancels out gravity. Hmmm, suppose I somehow reverse falling by thrusting upward. Can I then produce gravity? Aha!

To understand what Einstein had in mind, let us inflict an even more elaborate April Fools' joke on our friend. This time, after drugging her and putting her inside the box, we fly her deep into intergalactic space, far away from any gravitational field of force. Now rev up the engine and accelerate the whole contraption at a constant rate. When she wakes up, she notices nothing unusual at all. No floating sensation this time. She drops an earring, and it promptly falls to the floor (Figure 1.4). But to an outside observer, floating in space and watching the spaceship go zinging by, the dropped earring is actually floating in space in happy ignorance of the fact that the floor is rushing at it with ever-increasing speed. If we accelerate the rocket at precisely the right rate, our friend would see her earring falling to the floor exactly as always. By accelerating the rocket—in effect, reversing free-fall—we can "produce" gravity.

Clearly, if our friend had dropped her bracelet as well as her

1.4. *Another April Fools' prank. The dropped earring and bracelet "fall" at the same rate.*

earring, releasing them both simultaneously from the same height, she would observe them hitting the floor at precisely the same instant. But what is to her a mysterious universality is laughably obvious to the observer floating about outside: The floor is moving up to meet the floating earring and bracelet and so obviously arrives at the two objects at the same time.

A BALL OF WHISKEY

That a person inside an accelerating rocket ship feels an effect equivalent to gravity, an effect that disappears as soon as the ship stops accelerating, is illustrated wonderfully by the Belgian children's-book writer Hergé in his *Explorers on the Moon*, published in 1954. The story is part of a marvelous Belgian-French series of comic books describing the adventures of a redheaded boy named Tintin.

Two panels are shown in Figure 1.5a and 1.5b. On a trip to the moon, Captain Haddock, an alcoholic sea captain and Tintin's faithful friend, had sneaked some whiskey on board the spaceship, hiding the bottles inside a hollowed-out astronomy book. The first panel shows the captain about to enjoy his drink. When the spaceship was much more than four thousand miles away from the center of the earth, the effect of the earth's gravity became negligible, and any effect of gravity was due to the spaceship's acceleration. Since Hergé showed the characters going about their business as in a normal gravitational field, we can infer that the spaceship was accelerating at precisely the rate needed to reproduce the earth's gravitational field.

Just as the captain was about to set lip to glass, Thomson, a fumbling detective along for the trip, turned off the rocket engine, and the spaceship stopped accelerating. The whiskey, suddenly feeling no gravity, had no further compunction to stay inside the glass. The so-called

1.5a   **1.5b**

A ball of whiskey. In **1.5a**, *Captain Haddock has smuggled on board a spaceship a bottle of whiskey hidden inside a hollowed-out book on cosmology. Just as he is about to set lip to glass, a bumbling character named Thomson accidentally turns off the spaceship's engine. The spaceship stops accelerating. In the absence of gravity, the whiskey curls itself up into a ball and floats out of the glass. In* **1.5b**, *Tintin has managed to turn the engine back on. The spaceship accelerates and Captain Haddock and the ball of whiskey fall to the floor.*

surface tension, a force due to the attraction felt by the whiskey molecules for each other, curled the whiskey up into a ball, as Hergé correctly depicted. Eventually, Tintin managed to turned the engine back on, at which point the captain and the whiskey both fell crashing to the floor (or, as an outside observer would say, the accelerating floor rushed up to the captain and the whiskey). See Figure 1.5b.

Incidentally, had the captain taken the old man's toy along, he would have noticed the brass ball popping neatly into the cup just as his whiskey floated out of his glass.

That acceleration can produce effects identical to gravity has become quite familiar in this age of rocketry and fast elevators. Riding an elevator, we feel our internal organs sag momentarily in response to the additional gravitational pull produced as the elevator accelerates upward. For that moment, our stomachs have not yet realized that the elevator has taken off. Remarkably, this deceptively simple thought of the patent clerk, when suitably formulated into a principle of physics, will turn out to have far-reaching implications, which we will explore in the next chapter.

Before going on, let us summarize: (1) A falling person does not know she is falling because everything around her is falling at the same rate. (2) This apparently paradoxical thought has actually been verified by the experience of astronauts. To understand this, we have to understand in turn Newton's insight that orbiting space stations are actually in free-fall. They stay up because of their forward movement. (3) A person in an accelerating box is fooled into thinking that she is in a gravitational field, while in fact she could be light-years away from any gravitational field. To an outside observer floating by, however, a dropped bracelet is not falling to the floor. Rather, the floor is rushing up to meet the bracelet.

DESIRE TO FALL

While with her husband on the top floor of a skyscraper, [the patient] had an image of falling out of the window. The fantasy was so vivid that she shouted for help. When her husband questioned her, she realized that the fall was purely in her imagination. . . . Numerous clinical observations suggest that a person visualizing a scene may react as though it were actually occurring. . . . Even though the daydream may be temporarily experienced as reality (as a sleeping dreamer experiences a dream), anxious patients are able to regain their objectivity and label the phenomenon a fantasy.

—A. T. BECK et al.,
Anxiety Disorders and Phobias

Standing on the edge of a cliff facing the breaking waves, or leaning over the balcony of an upper floor of one of those towerlike contemporary hotels, looking down into the lobby, I have often mused about how a falling person feels no gravity. Surely, the patent clerk's daydream surfaced from the subconscious of an acrophobe.

Of all the phobias, acrophobia is perhaps the most common. At a conscious level, the acrophobe is afraid of falling. At the subconscious, he is afraid of his desire to go over the edge and fall. At least, that is how I feel—the urge to go into free-fall!

Falling—we have all had the experience of awaking suddenly from a dream in which we are falling into a dark void. Millions of years of experience instilled in our arboreal ancestors the fear of falling. Yet we cherish the experience: Children love the playground slide—it is the primordial fear momentarily mastered.

In the dream of falling, curiously enough, it is not entirely discomforting to be falling into a dark void. As long as we don't think there is solid ground at the bottom, we don't mind. We are terrified not by the fall itself but by the end of falling, by the sickening crunch of bone on ground. When we are no longer allowed by the ground to fall, that's when we feel gravity. But when we are falling, we are free from the grasp of gravity. It is the beckoning temptation of this freedom that terrifies the acrophobe.

By Einstein's own admission, he was daydreaming when what he called the happiest thought of his life occurred to him. Perhaps he was having the dream of falling. The subconscious can work in strange ways.

Hastening Through Space and Time

HA, HA, WE FOOLED YOU

Einstein was led by his daydream to enunciate the following principle: In a small enough region of space, the physical effects of a gravitational field as perceived by an observer are indistinguishable from the physical effects reported by another observer accelerating at a constant rate in the absence of a gravitational field.

This somewhat academic mouthful is known as the equivalence principle: It states that the two situations are physically equivalent. By using the setting of an April Fools' prank, I bring out clearly the meaning of the equivalence principle. When we accelerate our friend in a box furnished exactly as her living room, there is no way for her to tell, by performing any physical experiments and measurements inside the box, that she is no longer sitting in a gravitational field on earth.

In the equivalence principle, the rate of acceleration required depends on the strength of the gravitational field to be mocked up. If we want our friend to think that she is sitting in the strong gravitational field of a planet more massive than the earth, we would have to accelerate the box more.

I love the equivalence principle. It is the physics principle with the playful face. Ha, ha, we fooled you!

Perhaps I should say a word or two about the term *gravitational field*. Physicists introduced the concept when they got tired of speaking of the gravitational force between two objects. Instead, they think of a massive object, such as the earth, producing a field of force around it. When another object, be it the apple or the moon, is introduced into this field, it feels the gravitational force of the earth. The field turns out to be a useful concept because it allows us not to have to specify the object interacting gravitationally with the earth and, more important, because the field can be considered as a physical entity in its own right.

ILLUSION AND REALITY

In the precise statement of the equivalence principle, the phrase "in a small enough region" is what a lawyer might call fine print. Without this caveat, you may well object to the equivalence principle.

Suppose our friend is incredibly wealthy and her living room is so vast that from one end of the room to the other the earth has curved significantly. (See Figure 2.1.) When she is sitting on earth, she can see that two objects dropped at the two ends of the room do not fall parallel to each other. The two objects do not fall down; rather, they fall toward the center of the earth. She understands all this well, as she was taught that the earth is round. (For instance, suppose her living room is one mile long. Then, since the radius of the earth is about four thousand miles, she expects a one-part-in-four-thousand effect, or a 0.025 percent effect. For every foot the two objects fall, they get closer by 0.00025 foot. She could detect this effect with sufficiently accurate instruments.)

2.1a   2.1b

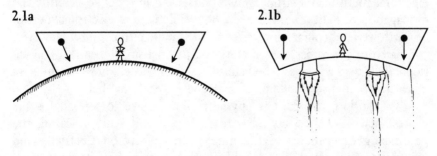

2.1. *The fine print that comes with the equivalence principle says that it works only in a small region. If our friend's living room is long enough, she would be able to tell that we are playing an April Fools' prank on her.* **2.1a.** *On earth, two objects dropped at the two ends of her living room would* not *fall parallel to each other.* **2.1b.** *In space, the two objects would* fall parallel to *each other.*

On the other hand, if we are accelerating our friend deep in space far from any gravitational field, she would not notice that two objects dropped at the two ends of the room do not fall in the same direction. (We cannot accelerate the two ends of the room in two different directions without tearing the room apart.) Thus, if her living room is large enough and her detection instruments sensitive enough, she would not be fooled.

The equivalence principle applies only in a region over which the gravitational field is uniform in magnitude and direction. In our example, our friend's living room must be small enough for this to hold.

Sometimes people are confused into thinking that the equivalence principle says that gravity equals acceleration. Rather, the equivalence principle is about illusion and reality. When the reality is that we are accelerating our friend deep in intergalactic space, she has the illusion that she is in a gravitational field. The illusion is equivalent to, but not equal to, the reality.

WORK SAVER

Einstein's insight makes physicists happy because it saves them a lot of work. Suppose, for instance, that they want to know the laws of electromagnetism in a strong gravitational field in order to describe the propagation of light around a black hole. The laws of electromagnetism on earth were, of course, discovered in the nineteenth century. However, compared to the forces of electricity and magnetism, the earth's gravitational field is so weak as to be utterly negligible. Well, how are the laws of electromagnetism modified in a strong gravitational field?

A priori, the curious physicists would have to set sail for the nearest black hole, set up a lab there, and spend years repeating the painstaking experiments of the nineteenth century on electricity and magnetism. But no, Einstein told us, they merely have to pack their instruments and a research assistant into a box, accelerate the box, and have the assistant report back on what she sees.

But there is no need to go to all this trouble—the physicists can lie back and simply imagine the whole exercise. The reason is that since the turn of the century, physicists have worked out how the descriptions of a physical phenomenon reported by observers moving relative to each other will differ. In other words, if I observe that a magnet next to a wire carrying an electric current behaves in a certain way, then I can calculate what a person whizzing by me will observe. According to the equivalence principle, what the person whizzing by observes is in fact what a person in a gravitational field will observe.

Thus, once we master a physical law in the absence of gravity, be it the phenomenological law governing the flow of water or the more basic law governing the behavior of some subnuclear particle, we can immediately find out what the law is in the presence of gravity by appealing to the equivalence principle. In the next few sections, we will see how easily this procedure works in practice.

THE BENDING OF LIGHT

Does light fall? Newton and many of his contemporaries thought so, since they imagined light to be made up of tiny little balls. Physicists after them were not so sure. After all, you can't quite hold light in your hand and feel its heft.

Einstein settled the question decisively, thus illustrating beautifully the power of the equivalence principle. To see how, let us go back to our friend, blissfully unaware that we have been accelerating her and her entire living room in deep space. She decides to amuse herself with a game of laser tag. She fires, and a flash of light zings across the room. But the floor, totally indifferent to what she might be doing, rushes happily upward as always and so gets closer and closer to the flash of light. Our friend sees to her astonishment that the beam of light is bending toward the floor.

Ha, ha, we say with a laugh, we fooled her! Light is not falling, but the floor is rushing upward! But according to the equivalence principle, the physics inside her living room accelerating deep in space is the same as the physics in a gravitational field. We conclude, and rightly so, that gravity bends light.

Thus, in one clean swoop, the equivalence principle settles a long-standing debate. Observations have since proved that the conclusion is correct: Starlight passing by the sun is indeed bent slightly.

ANTIGRAVITY

Can there be antigravity? I once saw a declassified government document studying the capability of enemy nations to produce an anti-gravity warplane. (It is quite touching, really. The document identifies a number of Soviet-bloc physicists working on the theory of gravity. I know many of them to be totally theoretical types who may be hard-pressed to explain how even a regular airplane would work.)

Science fiction writers love the notion of antigravity. But if you believe in the equivalence principle, then antigravity is impossible.

The argument is again ludicrously simple. Drop a bunch of objects. Can any of them fall upward? No. According to the equivalence principle, you might as well be inside a rocket ship accelerating in deep space. From this point on, it is the same old argument. To an observer floating outside, the floor is rushing up to meet the objects you dropped, even though you could swear they are falling toward the floor. Obviously if the equivalence principle is correct, then in a gravitational field also, there can't be an object that falls upward.

TIME WARP

From the equivalence principle flows an astonishing array of effects beyond human dreams. Perhaps none has gripped the public imagination more than the warping of space and time by gravity. But yet Einstein's happy thought, at once deceptively simple and unfathomably profound, almost immediately implies the warping of space and time.

Picture two observers at two different points in a gravitational field, on the ground and on top of a tower, comparing the apparently immutable flow of time. The ground observer agrees to signal the passage of time by sending the tower observer a pulse of light every second.

According to Einstein, it makes absolutely no difference if we imagine the entire arrangement, the ground with the ground observer sitting on it and the tower with the tower observer perched on top, in an accelerating spaceship deep in space. It is now easy to see qualitatively what happens. As a given light pulse approaches the tower observer, the tower observer, who has been accelerating ever since the pulse left the ground observer, is moving away faster and faster. Thus it takes an extra amount of time for the light pulse to reach the tower observer. Time appears to flow at different rates for the two observers. The equivalence principle asserts that back on earth, the observer on the ground and the observer on the tower, even though both are sitting perfectly still, will see the same disparity in time flow.

That is all time warp is, dear reader, astonishing because it is fact, but not as bizarre as science fiction writers would like it to be.

SPACE WARP

To understand why gravity curves space, consider for a moment how we could possibly determine whether the three-dimensional space we live in is curved or not. Since curved three-dimensional space is hard to visualize, let us consider instead how a tiny two-dimensional creature living in a curved two-dimensional space—that is, a curved surface—could possibly determine whether his space is curved or not.

When we think of a curved surface, we naturally think of it as embedded or contained in the three-dimensional space of our everyday experience. As three-dimensional beings looking down at a two-dimensional world, we can easily see whether the surface is curved. But the creature cannot get out into the three-dimensional space and look at his world any more than we can get out of our universe to take a look. What can the creature do?

The key is provided by a hoary riddle that puzzled me as a kid. A hunter walks due south for a mile, then turns due east and walks for another mile. Finally, he turns due north. After walking for a mile, he arrives back at his starting point, where he encounters and shoots a bear. What color is the bear?

The information provided in the riddle indicates that the surface of the earth must be curved. To a mathematician, an unmistakable sign is that the three angles in the triangle formed by the hunter's trek add up to $90° + 90° + 90° = 270°$, which is more than $180°$. (See Figure 2.2.) In contrast, the three angles in a triangle drawn on a flat space always add up to exactly $180°$. Thus our two-dimensional creature only has to draw a triangle in his space and add up the three angles contained in it to see if the sum is $180°$ or not.

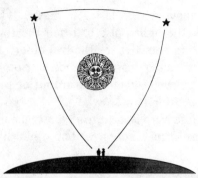

2.2. *An astronomer observing the light from two distant stars. As the light passes by the sun, they are pulled toward the sun. Einstein thus concluded that space is curved. The figure is not drawn to scale, of course.*

Well, we can do the same thing. Pick two stars and consider the triangle with us on earth at one point and the two stars in the sky at the other points. (The reason for the stars is that we want the triangle to be large. If you just walk around your neighborhood, the earth looks perfectly flat.) Add up the angles. But wait—how do we draw the straight lines that form the three sides of the triangle? It is of course important that the sides of the triangle be straight. Notice that in the riddle the hunter is always walking in a straight line.

The only physical way we know to define a straight line is by the blazing path of a beam of light. Indeed, that is what surveyors use. There is no alternative. Thus defined, a straight line provides the shortest distance between two points.

Now comes the punch line. The equivalence principle tells us that gravity bends light. Thus the angles in the triangle formed by us and the stars will exceed 180°. (See Figure 2.2.) And thus Einstein concluded that space itself is curved, in the same way our imaginary two-dimensional creature concludes that his world is curved.

THE CURVED UNIVERSE

An apparently innocuous principle coming out of the old man's toy implies that the entire universe is curved. Loosely speaking, a curved surface can curve inward or outward on itself. The surface of a sphere curves inward and eventually closes on itself. In contrast, something like the surface of a saddle curves outward and if continued would extend out to infinity. (See Figure 2.3a.) How can a creature living on the surface tell whether the surface curves inward or outward? Once again, he forms a triangle. As Figure 2.3a suggests, the angles of the triangle will add up to less than 180° if the surface curves outward.

Similarly, our universe could turn out to be either closed, like the surface of a sphere, or open, like the surface of a saddle. If closed, the universe would be finite. If open, the universe would be infinite. But note that in either case the universe would not have an edge.

The curvature of the universe is determined by the matter in it. Which way the universe actually curves has to be decided by observation. More on this later.

A friend of mine (a real-life one, not the one we sent off into deep space) complained to me that even though she has heard of a curved

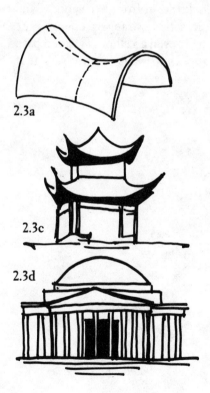

2.3a

2.3c

2.3d

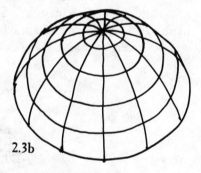

2.3b

*The concept of open and closed spaces is illustrated here by curved surfaces. **2.3a.** An open surface looks something like a saddle. **2.3b.** A closed surface looks something like a portion of a sphere. The notion is that if a closed surface is extended, it tends to close upon itself. The concept of open and closed surfaces also appears in architecture, as in these two roof designs: in **2.3c**, an open roof and in **2.3d**, a closed roof.*

universe several times, she just can't picture it. Nobody can, really. We can picture curved two-dimensional space easily enough, but ask the tiny denizen of two-dimensional space to do it, and he can't.

RAPUNZEL'S VISION

The curvature of space and time lies at the heart of Einstein's understanding of gravity. A gravitational field curves space and time; conversely, the curvature of space and time signals the presence of a gravitational field. In a very real sense, the curvature of space and time *is* the gravitational field. To explain this identification, imagine perpetrating yet another April Fools' joke, meaner and more elaborate than ever.

Raise our friend (if she can still be called our friend after all this), confined to a room high on top of a tower, Rapunzel style, and strictly isolated from any curved surfaces (Figure 2.4). From birth, she has seen only flat, plane surfaces. One day she looks down from her window and sees children playing marbles. She notices that, strangely enough, the marbles follow curved paths and occasionally even an elliptical closed path around a certain point. The marbles appear to want to approach that

2.4. *Rapunzel's vision.*

point, and they accelerate as they get near. To explain her observations, our friend soon concludes that there is a mysterious force of attraction between the marbles and some object she can't quite see located at that point.

We then let our friend out of the tower and reveal that the ground where the children play marbles is not quite flat but that there is a depression at the mysterious point of attraction. The force that she saw from high on the tower came about because of the depression. Just like our friend, we at once thought of a mysterious gravitational force of attraction between the sun and the planets, causing the planets to follow paths much like the marbles in our tale.

What is gravity? That age-old question was finally answered: Gravity is curved space and time. We now understand gravity as a sort of "mattress effect": A person lying down on a soft mattress creates a depression, toward which another person would tend to roll. Much in the same way, a massive object curves the space and time around it, and a neighboring object, feeling the curvature, would tend toward that massive object.

With this identification of gravity as curved space and time, we now understand that the planets are merely trying their best to follow the path of least distance in the curved space and time around the sun, much as mariners trying to navigate the path of least distance on the curved surface of the earth follow the great circle route.

The mysterious universality of gravity now becomes clear: All objects respond the same way to a gravitational field for the simple reason that they are all following the path of least distance in curved space and time. They are all travelers in a mysterious landscape not of their own making, and as fellow travelers, they must all follow the same path.

Analogously, if a ship's skipper wants to sail from San Francisco to Hawaii over the shortest distance possible, he would have to follow the same path be his ship a frigate or a pleasure craft. The path depends on properties of the earth's curvature, not of the ship.

As you sit in your armchair and read these words, you may not think of yourself as falling, but of course you are. You are attached to the earth, and the earth is falling toward the sun. (Well, as Newton explained, the earth also has some forward momentum and thus it never quite gets to the sun. At least not for a while. Remember how as an orbiting shuttle falls, the surface of the earth keeps "dropping" away.) It may seem very strange—and it *is* strange—that as you sit quietly, you and the earth are in fact rushing through space and time cruising the path of least distance.

Hastening through the utter emptiness of space and time, all material objects cruise on. Silently and unerringly, they reach their destinations without ever traveling more distance than necessary.

A NEW THEORY OF GRAVITY

Thus far we have talked only about how objects respond to a given gravitational field. We have yet to discuss how gravitational fields are produced. It is as if we have described how an individual is affected by the fluctuations in the stock market, but we have yet to understand how the fluctuations are produced. Thus far Einstein had half of a theory of gravity.

He was to spend ten years searching for the other half. The equivalence principle again turned out to be the key.

Think of the equivalence principle as a demand or restriction on physics. If someone comes along and tells you that an electromagnetic phenomenon in the presence of gravity must be the same as that phenomenon in the absence of gravity but in an accelerating box, he is putting a restriction on any possible theory you can invent to describe the interplay between electromagnetism and gravity.

Oh, you say you've got a theory. Well, let's check if it satisfies the equivalence principle. No, it doesn't! Into the wastebasket it goes.

Take Newton's theory. Old Newton said that all massive objects, those very same objects that respond to gravitational fields, also produce gravitational fields. The gravitational fields produced by massive objects act in turn on other massive objects and influence their movements. Our analogy about the stock market is apt. The fluctuations in the stock market are produced by individual investors either acting on their own or through their representatives in such financial institutions as pension funds. These investors are the very same individuals who are affected by these fluctuations. The more dough you have, the more effect you can have on the stock market. Similarly, the more massive an object, the stronger the gravitational field it produces.

Fine. That's what Newton said. As is worthy of his name as a physicist, he also wrote down an equation saying precisely how strong a gravitational field is produced by a given mass. Now we come to the sixty-four-thousand-dollar question: Does Newton's equation satisfy the restriction imposed by the equivalence principle?

No, it doesn't!

To see that it doesn't, we really have to look at the equation. To understand this roughly, however, we can again think of the bending of light by gravitational fields. Newton wasn't sure whether light consists of tiny massive balls, but we now know that the particles of light, the photons, do not have mass. Since the gravitational force acting on an object is proportional to its mass, in Newton's theory, light does not bend in gravitational fields. The equivalence principle, however, predicts the bending of light in gravitational fields, and thus it couldn't possibly be satisfied by Newton's theory.

Einstein, with his total faith in his happy thought—the equivalence principle—was thus forced to modify Newton's equation to satisfy the restriction imposed by that principle. That's the hard part. It took ten years before he found the right way to do it.

Although the mathematical subtleties of the theory confounded the great man himself, the bottom line is not hard to understand. The theory tells us how much mass is required to produce space and time of a desired curvature. You want it more curved? Just put in more mass.

SYMMETRY AS RESTRICTION

So, the equivalence principle imposes a restriction on the possible theories of physics. Roughly speaking, they must have a certain form.

In everyday life, we tend not to think of the aesthetic principle of symmetry in terms of restrictions. But symmetry does imply restrictions.

Suppose you were to tell a schoolchild to draw a pattern with threefold symmetry under rotation. (I don't know about you, but when I was a child, there was one year when I had to draw lots of patterns with specified symmetries.) A child who turns in the Mercedes-Benz symbol gets high marks, but not the one who draws the Volkswagen symbol. The Mercedes-Benz symbol is threefold symmetry—if you rotate it through 120 degrees, one third of 360 degrees, it looks the same. The Volkswagen symbol, in contrast, is not threefold symmetric. (See Figures 2.5a, 2.5b, 2.5c.)

By telling the child to draw a pattern with a threefold symmetry under rotation, you are effectively restricting what the child can draw. Thus, symmetry imposes a restriction. Symmetry tells us what can and what cannot be.

Physicists in their search for the ultimate theory love symmetries for an obvious reason: Symmetries narrow down the choices. The search for beauty and design in Nature is tantamount to a search for symmetries. My previous book, *Fearful Symmetry*, tells the stirring story of this search.

The equivalence principle conceived as a symmetry was crucial in guiding Einstein to his theory of gravity. The point is that symmetry gave him the theory in its entirety in one fell swoop. Had he tried to patch up Newton's theory by bits and pieces, he wouldn't have gotten very far.

In our analogy, the child who drew the Volkswagen symbol then catches on to what the teacher wants. But if he were to lengthen a line here and shorten a line there, he would have a hard time reaching a threefold symmetric pattern. On the other hand, once he understands what the heck "threefold symmetry under rotation" means, he can then change his pattern easily. For instance, he can erase the W and put in two more V's at the appropriate places.

Our analogy would have been even better if this pattern turns out to be so singularly beautiful that everyone who sees it swoons and says with a sigh, "That's it! That's the only pattern with threefold symmetry worthy

2.5a   2.5b   2.5c

The figure in 2.5b is not threefold symmetric under rotation around its center, while the figures in 2.5a and 2.5c are. The figure in 2.5c is obtained from the figure in 2.5b by erasing the W and putting in two Vs at the appropriate places.

of our aesthetic sensibility!" In physics, that actually happened. Upon first seeing Einstein's theory of gravity, physicists tend to swoon.

THE DYNAMIC UNIVERSE

With his theory of gravity constructed, Einstein could now contemplate the universe itself. Not only did he understand how stars and galaxies move around in the curved universe, but he also was able to calculate how the masses contained in the universe curve space and time, the fabric of the universe. A new cosmology was at hand.

While the stars and galaxies cavort about, would space and time stand still and not respond? Hearing the same beat, might not space itself flex and twist, expand and contract? As we will see, Einstein missed the opportunity to make perhaps the most exciting prediction in the history of physics—that the fabric of the universe is dynamic.

By now the news has long been out that the universe is expanding. Later, we will tell how that astounding discovery was made—through the sheer will of the human race to look outward.

From the old man's toy to the secret dynamics of the universe. The range of the human intellect elates and astonishes.

3

The Mighty Shall Be Weak

THE MIGHTY HAND OF GRAVITY

When my son Andrew was about four years old, he asked me whether gravity or Superman were stronger. Clearly, I replied, Superman's got to be stronger, since he can zip away from the earth's gravity at a moment's notice. But on second thought, is Superman strong enough to move galaxies around? Can Superman rein in the expansion of the whole universe? Only gravity is strong enough to get the universe to perform.

At this very moment, our galaxy is hurtling toward the Virgo cluster of galaxies at a couple of hundred kilometers a second, faster literally than a speeding bullet. Virgo is pulling us into its gravitational field. At the same time, our galaxy continues to rotate sedately: The invisible hand of gravity ties together the 100 billion stars that make up the galaxy. Our dear old sun stays within this hurtling swarm of stars only because of its attraction to all those other stars, just as our earth stays within the solar system because of its attraction to the sun. On an even smaller scale, we stay tied to the earth by gravity, with nary a fear of falling off. On a grander scale, as the universe tries to expand to ever larger size (what an egomaniac, this universe!), every bit of matter it contains is trying to rein the universe in by pulling gravitationally on every other bit of matter. If you were half as mighty as gravity, you would be incredibly famous.

BY AN ABSURD MARGIN

And yet, as physicists began to understand forces early in this century, they discovered that mighty gravity is by an absurd margin the most feeble force in Nature.

Compare the electromagnetic force with the gravitational force. No contest. The electric attraction between two protons is some 10^{36} (the number 1 followed by thirty-six zeroes) times stronger than their gravitational attraction. As humongous numbers go, 10^{36} is awfully humongous. Even as a physicist, I am numbed by these enormous numbers. What does it really mean to say one force is 10^{36} times stronger than the others?

To put the contest in somewhat more vivid perspective, picture two protons one centimeter apart. Then picture another pair of protons separated by one light-year. (A light-year is, of course, the distance traveled by light in one year. The relevant figure to remember is that in one second light travels a distance equal to about seven times the circumference of the earth.) Picture the two pairs of protons. The electric force between the pair separated by one light-year is equal in strength to the gravitational force between the pair separated by one centimeter. That is how feeble gravity is compared to electromagnetism.

TOTALLY NONDISCRIMINATING

Gravity, however, makes up for its weakness in a striking manner: It acts on all particles, be they rich or poor, black or white, massive or massless, positively charged, negatively charged, or not charged at all. The gravitational force between any two particles is always attractive. As you read this, every proton, electron, and neutron in your body is attracted to every particle in the earth. There is a rock in Madagascar. Every proton, electron, and neutron in that rock is pulling you to it. And the same goes for all the particles in the molten magma in the center of the earth, just as Newton thought.

In contrast, the electric force acts only on charged particles, and furthermore acts on positively charged and on negatively charged particles in opposite ways. A lump of macroscopic matter such as your body or the earth is composed of positively charged protons, negatively charged electrons, and uncharged neutrons. The number of protons is almost exactly equal to the number of electrons so that macroscopic objects are essentially electrically neutral. The electric force between macroscopic objects cancels to zero to a high accuracy. The attraction a proton in your body feels for all the electrons in the earth is balanced almost exactly by the repulsion it feels for all the protons in the earth.

Indeed, it is precisely because the electric force is so strong that macroscopic objects are normally electrically neutral. If one object contains an excess of electrons while another is deficient in electrons, the attraction of the protons in the second object for the excess electrons in the first object is so strong that the electrons would jump across thin air to get from one object to the other. Witness the fury of lightning storms.

Gravity, on the other hand, is totally nondiscriminating. It is only because macroscopic objects contain such an enormous number of particles that gravity makes its presence known. Every single particle in your body is pulled toward every single particle in the earth, and the effect adds up. It takes a huge object like the earth to exert a significant pull on us.

You can turn it around and say that a force that acts on every particle had better be extremely weak, given the enormous number of particles abroad in the universe. Gravity could not be much stronger without the universe's looking completely different.

That gravity acts on all particles can ultimately be traced back to the equivalence principle, the happy thought of the patent clerk. Since the floor is coming up to meet the falling objects, all objects necessarily fall in the same way.

ADEQUACY OF NEWTONIAN COSMOLOGY

Because of the extreme weakness of gravity, the differences between Newton's theory and Einstein's theory are minute and difficult to detect. In all those cases where the differences could be detected, Einstein's theory was verified over Newton's.

Newton's theory, however, continues to be enormously useful.

While those physicists intent on differentiating Einstein's theory from Newton's have to strain their ingenuities, their colleagues who have to calculate the effects of gravity see the minuscule difference between the two theories as a tremendous blessing. In almost all situations, it is adequate to use the much simpler Newtonian theory. For instance, even right at the surface of the sun, space is curved only by one part in five hundred thousand. Thus the vast space between stars is very flat indeed. Even when cosmologists talk of swarms of galaxies, they can simply use Newtonian physics.

Einstein's theory is essential, however, when we want to discuss the universe as a whole. As an analogy, the flat-earth theory and the round-earth theory differ negligibly over a distance of a few miles. But as one travels over longer and longer distances, the differences between the two theories become more and more noticeable until finally, when one has circumnavigated the globe, the flat-earth theory fails totally. Similarly, the differences between Newton's and Einstein's theories become significant when we consider the universe as a whole. In particular, in Einstein's theory the universe may be closed like the surface of a globe.

UNAVOIDABLE COLLAPSE

Since it is so difficult to compare the two gravitational theories by studying the cumulative effects of a weak gravitational field built up over vast distances, the alternative is to study strong gravitational fields, such as those surrounding black holes. The idea of the black hole is neither new nor particularly profound. As early as 1795, the Marquis de Laplace remarked that even light may not move fast enough to escape an extremely dense astronomical object. The dense object pulls the light in.

The crucial question is how black holes could have formed. One possibility is that they were formed from the gravitational collapse of extremely massive stars that had burned up all their nuclear fuel. The parts of the collapsing star pull each other in toward the center.

It is in the formation of black holes that Einstein's theory differs from Newton's. To understand why, we have to discuss Einstein's other great discovery, that mass and energy are effectively the same. In Newtonian physics, mass is mass and energy is energy. A moving object has energy associated with its motion, but an object at rest contains no energy. Einstein, however, discovered that an object of mass m sitting at rest contains an amount of energy given by $E = mc^2$. He showed that mass can be converted into energy and energy into mass.

If mass and energy are effectively the same, then mass and energy can both produce gravity. In contrast, in Newton's theory, gravity is produced exclusively by mass.

Let us go back to the collapsing star. As the matter in the collapsing star is squeezed, it becomes stiffer to resist further collapse. The star's collapse would eventually be arrested.

This Newtonian story is incomplete, however. The point is that energy is necessarily associated with the stiffness (think of the energy put into a compressed spring waiting to be released). In Newton's theory, this energy does not generate any gravitational field. In contrast, Einstein's theory states that the energy contained in the compressed star, being equivalent to mass, does generate an additional gravitational field, which in turn would hasten the collapse by giving an additional inward tug to the matter. By stiffening up, the matter only manages to strengthen the inward crush of gravity. You cannot avert, even in principle, the impending formation of a black hole.

LIKE COMPOUND INTEREST

That energy as well as mass can generate a gravitational field underlines the fundamental difference between the two theories of gravity. In Newton's theory, a massive object generates a gravitational field around it, and that is the end of the story. The situation with Einstein is considerably more complicated. The gravitational field around the massive object contains energy and thus generates an additional gravitational field. The star, in other words, generates a gravitational field, which in turn generates an additional field, which generates yet another field, and so on ad infinitum. It is the gravitational analog of money begetting money through compound interest.

This piling of field on top of field makes Einstein's theory much more complicated mathematically than Newton's, much as compound interest is harder to compute than simple interest.

Normally, the additional gravitational fields generated are small, and Einstein's theory differs little from Newton's. Around a soon-to-be-born black hole, however, the additional gravitational fields pile up, causing an extreme warping of space and time, until, roughly speaking, the fabric of space and time itself tears. The collapsing star is literally crushed out of existence.

Not even Superman can escape from the clutch of a black hole once he starts falling in. Now you know who is stronger, Superman or gravity.

JELL-O WAVES

Because gravity is so feeble, one key consequence of Einstein's theory of gravity remains unverified directly to this very day.

In Einstein's theory, gravity is the manifestation of curved space-time. Space and time can be squeezed and stretched. The Soviet physicist and humanitarian Andrei Sakharov once likened space to an elastic medium that bends and warps in response to the presence of matter, much like a piece of Jell-O in the hands of a three-year-old. One characteristic of elastic media is that waves can propagate in them. Poke a piece of Jell-O, and it will shake and roll in indignation. Imagine a very long piece of Jell-O. By poking and tapping the Jell-O at one end, you can send a wave of Jell-O vibration traveling down to the other end. I suppose you can communicate with a friend that way.

Similarly, waves can propagate in the elastic medium that is space. Two galaxies collide in silent fury, and space in the vicinity goes into a dance of frenzied contortion. It gets warped and curved and twisted. Because space is elastic, it tries to bounce back. This warping and unwarping gets communicated to the region of space nearby, which in turn gets warped, and thus all this contortion propagates outward in a wave of disturbance. As another analogy, recall that we likened the gravitational field to a soft mattress. If the persons lying on the mattress were to thrash about, waves would be set up in the mattress. Thus Einstein's theory predicts the existence of gravity waves. Indeed, any theory that describes the gravitational field as a dynamic entity would predict the same.

How would we know when a gravity wave has come by? Well, how would a bacterium on the Jell-O know that a Jell-O wave has come by? As the wave comes by, the Jell-O shakes and rolls and the bacterium moves accordingly. In general, this is how you would detect any kind of wave. Sitting in a small boat on a calm day, when you bob up and down you know that a wave from a passing speedboat has just come by. When a light wave reaches your eyes, the electrons in your retina bob up and down. When a radio wave reaches your radio, the electrons in the radio's circuits bob up and down. And so on.

Similarly, when a gravitational wave comes by, space will warp and unwarp and things will move in response. Sitting on a small planet on a calm day, you know that a gravity wave has come by when you and your planet bob up and down.

Yes indeed, a violent gravity wave would throw planets around helter-skelter, and if you are prone to worry about such things, you can imagine terribly nasty doomsday scenarios. Happily, physicists expect that when a gravity wave does come by, far from being violent enough to throw planets around, it will be so gentle as to be barely noticeable by the most sensitive instruments present-day technology can build.

The reason is, once again, the extreme weakness of gravity. In addition, any credible sources of gravity waves are separated from us by the vast distances astronomers are so proud of. Out there, galaxies could crash into each other and black holes could suck whole civilizations up, and we would hardly notice the disturbances. It would be like detecting the wave generated by a passing speedboat a hundred miles away.

GRAVITY WAVES EXIST

Physicists do not doubt that gravity waves exist, since their existence follows from general considerations rather than from the details of Einstein's theory. Essentially, as soon as we grant that one object can affect another object gravitationally, we are committed to gravity waves. By saying that objects act on each other gravitationally, we say that when one object moves, the other will eventually know about it. To communicate the movement of one object to another, a carrier is needed for the signal, and that carrier is provided by a gravity wave. Actually, the whole thing is exceedingly simple. We feel the gravitational effects of distant galaxies. Thus, when distant galaxies collide, we will know about it. To say that a gravity wave can propagate across vast distances is the same as saying that gravity is long-ranged.

The same general considerations that demand the existence of gravity waves also demand the existence of electromagnetic waves. The electromagnetic force is long-ranged as the gravitational force is long-ranged, and the same argument—that the motion of a charge has to be communicated somehow to a distant charge—establishes the existence of electromagnetic waves. This is also why the speed of gravity waves is the same as the speed of light, light being a form of electromagnetic wave. The arguments, and the mathematics, parallel each other in the two cases.

TO DETECT RIPPLES IN SPACE-TIME

To be in the business of detecting gravity waves, you need enormous dedication and fortitude. Compare the frustration of the would-be gravity-wave catchers with the electromagnetic-wave catchers late in the last century. With the technology of the time, it was no mean feat, of course, that the pioneers of radio managed to detect radio waves. Nevertheless, it was significantly easier. The essential difference is that the physicists were able to generate electromagnetic waves in the laboratory. In 1888, Heinrich Hertz generated the first radio waves by passing sparks between two brass balls and detected the waves at the other end of the room. His detector was simple: two tiny brass balls separated by a small gap. Placing his eyes right up to his detector, Hertz could see a tiny spark jump across. Within years, the age of telecommunication was born.

In contrast, there is no conceivable way of generating gravity waves in the laboratory. Physicists can detect only what gravity waves the universe produces. There is no lack of sources. Indeed, in principle, the gravity waves emitted during Creation should still be around, waiting to be detected. By all accounts, the universe should be as noisy in gravity waves as in electromagnetic waves.

In fact, we have convincing, though indirect, evidence that gravity waves exist. Binary star systems, in which one star orbits around another, are fairly common in the universe. As the stars go round and round each other, we expect that they emit gravity waves and thus lose energy. As the stars lose energy, they spiral in toward each other. As luck would have it, some years ago a binary system in which one of the stars is a pulsar was discovered. The pulsar, with its regular pulse, provides a highly precise natural clock by which observers could determine how long it takes for the stars to complete one orbit. The rate at which the orbital period is changing agrees exactly with what is predicted by calculations using Einstein's theory.

For decades, physicists have built ever more sensitive detectors to look for gravity waves from astrophysical sources. So far, nothing. Obviously, in designing the detectors it is important to estimate, given what we know about the universe, how many gravity waves the universe has been emitting. The fact that gravity waves have not yet been detected may mean only that the detectors built to date are not sensitive enough. Unfortunately, the estimates cannot be absolutely reliable because our knowledge of the central cores of galaxies, for example, is not that firm.

At the moment, researchers from the California Institute of Technology and the Massachusetts Institute of Technology have jointly asked the U.S. government to fund an ultrasensitive detector that should be able to pick up gravity waves if the current estimate of how many gravity waves are coming in is correct. The detector consists of an L-shaped four-kilometer-long vacuum tube with a heavy mass suspended at each end. A laser light is bounced back and forth off mirrors mounted on the two masses to monitor the distances between the two masses. When a gravity wave from an astrophysical source passes by, the distance between the two masses is expected to change by 4×10^{-16} centimeters—about one thousandth the size of an atomic nucleus! How can you possibly measure that kind of distance change? You may gasp. The clever experimenters have come up with a scheme in which the laser light is bounced back and forth many times and thus amplifies the distance shift. Compare that with Hertz's eyeball! The experimenters say that if the National Science Foundation approves the project, the detector can be operating by 1991. To be sure that they have actually detected a gravity wave rather than just some local disturbance, the experimenters are asking for two detectors, to be located in California and in Maine, so that any signal picked up by one detector can be checked with the other detector. Eventually, with detectors located in different parts of the world, experimenters will be able to pinpoint the incoming direction of any gravity wave detected.

Meanwhile, people who have built detectors to look for gravity waves wait and hope. Of course, in principle, an unbelievably cataclysmic event of a violence beyond our wildest imaginings (and unaccountable for by any known physical processes) could have happened somewhere far, far away and generated a tsunami of a gravity wave approaching us at this very moment. There can be no warning whatsoever, since gravity waves travel at the speed of light. By the time we see with our telescopes something bizarre happening out there, the gravity wave would have arrived.

Well, if you like to lose sleep nights worrying about disasters, this is a good one to think about. From the invisible darkness of space, the hand of a gravity wave suddenly reaches out and flings the earth out of its orbit. The insignificant mites that we are will be thrown into airless space. We will choke to death if the escaping air from our innards does not tear us apart into smithereens first. Come to think of it, we will probably all die from fright. On the other hand, a less violent gravity wave will probably just send real-estate values plummeting.

ANOTHER WINDOW TO THE COSMOS

Quite aside from our human urge to know and to improve our understanding of gravity, physicists want to detect gravity waves because they promise to open up another window to the world out there.

For eons, our knowledge of the cosmos has come to us in the form of light. And then physicists discovered in the nineteenth century that light is only one form of electromagnetic wave. With the development of detectors for the other forms of electromagnetic waves, microwave astronomy, radio astronomy, infrared astronomy, ultraviolet astronomy, X-ray astronomy, and gamma-ray astronomy were born one after another. After all, astronomical bodies are not going to radiate electromagnetic waves only in those frequencies detectable by certain creatures on a particular speck of a planet. The universe is humming across the entire electromagnetic spectrum. It is as if we had been peering at the cosmos through a narrow window and all of a sudden the curtain was pulled back to reveal that the window was in fact quite wide.

Still, wide as the electromagnetic window is, it would be wonderful if we could look through another window. The frustration is that we know for sure that another window exists, the gravity wave window, but the signal coming through is too weak. If we ever do detect a signal, it will open up a fabulous new epoch in the human exploration of the cosmos. To have gravity-wave astronomy would be akin to our collectively growing a second set of eyes, but even better, since new types of signals will be received. As we will see in a later chapter, the universe may in fact be filled in the main by a form of previously unknown matter that does not radiate electromagnetic waves. Thus, what we have been seeing of the universe may be only a small part. The coming of gravity-wave astronomy will reveal the universe as we have never seen it before.

I am reminded of those sound and light shows that Europeans are particularly fond of. The universe is putting on a sound and light show also—more accurately a gravity- and electromagnetic-wave show. So far, it has been like a silent movie.

A HERD OF GRAVITONS

Thus far, our discussion of gravity has been based entirely on classical physics. At a deeper level, Nature is described by quantum laws.

Before going back to gravity waves, let us first review the more familiar case of electromagnetic waves. In classical physics, a light wave is simply a wave of electromagnetic energy. In quantum physics, however, energy comes in packaged units. When we examine a light wave more closely, we see that the wave actually consists of a huge number of tiny packets of electromagnetic energy called photons. In short, photons can be thought of as the fundamental particles of light.

The situation reminds me of those nature films with aerial shots of migrating herds of wildebeests. From a distance, we see a dark brown tide surging forward. As the lens zooms in, we see the tide differentiating into individual wildebeests thundering along. As we zoom in and examine Nature more closely, we see what classical physicists took to be a wave of light differentiating into individual photons cruising along. (This picture is somewhat oversimplified—one of the mysteries of quantum physics is that a photon can also be described as composed of waves. The picture is, however, adequate for our purposes here.)

Similarly, at the quantum level, a gravity wave consists of packets of gravitational energy called, appropriately enough, gravitons. Physicists have not even detected gravity waves yet, let alone a graviton. Indeed, to the extent that the future is foreseeable, physicists see no prospect of ever detecting individual gravitons. Nevertheless, as much as they believe in quantum physics, physicists believe in the graviton.

Classical physicists speak of massive objects responding to the gravitational fields generated by each other. To a quantum physicist, the gravitational field consists of a swarm of gravitons. A massive object generating a gravitational field is actually emitting these teensy-weensy bits of gravitational energy, while a massive object responding to a gravitational field is absorbing them. Quantum physicists analyze the gravitational interaction between two massive objects as the net effect of a fundamental process in which one object emits a graviton that is then absorbed by the other object. The process repeats itself rapidly. This constant exchange of gravitons between the two objects produces the observed gravitational force. Similarly, the constant exchange of photons between two charged particles produces the observed electromagnetic force between them. I liken this constant exchange of gravitons to the marriage brokers of old traveling between two parties, telling each the other's intentions.

Since the early days of physics, the notion of force has been among the most basic and the most mysterious. It was thus with considerable satisfaction that physicists finally understood the origin of force as being due to the quantum exchange of mediator particles such as the graviton and the photon.

ASHES, ASHES, WE ALL FALL DOWN

A child asks: Why do we fall down? A panel of experts replies.

Aristotle: Well, the earth is the natural home for rocks and men. Rocks fall because they want to go home. As rocks fall, they go faster and faster, much as recalcitrant rental horses would break into a gallop as they approach the stable.

Newton: That Aristotle fellow is full of it. I have interviewed plenty of rocks, and they never said anything about going home. Rocks and apples fall because they and the earth and every other object in the universe exert a force on each other. As a child jumps out of a jungle gym, the child is actually also pulling the earth up.

Einstein: Newton was so right, but there is more to the story. The force Newton talked about results from the curvature of space and time. The earth warps the space and time around the jungle gym so that when a child jumps, the child is merely following the natural curved path in that space.

The quantum theorist of gravity: Einstein always hated the quantum world. Otherwise, he might have realized that his curved space and time is due to gadzillions of gravitons sashaying around. As a child jumps out of the jungle gym, gravitons zing back and forth like crazy between the child and the earth.

Leave Aristotle aside—I really don't think what he said is right.

43

Then each new picture is built on the preceding picture. Such is the orderly progression of science. The quantum theorist of gravity does not say that Einstein was wrong, but merely that he did not include the effects of the quantum. Each picture reduces to the preceding one in a well-defined approximation. For instance, Einstein's theory reduces to precisely Newton's theory under specific circumstances in which it can be proved that certain effects are small enough to be neglected. In contrast, there is no sense in which Newton's theory reduces to Aristotle's theory. With each succeeding picture, we gain a deeper understanding. For instance, you may justifiably say that Newton did not explain gravity but merely described it. Why does the gravitational force between two objects decrease as the square of the distance between them? Why is the force proportional to mass? Such questions are beyond Newtonian theory. In Einstein's theory, however, the way the gravitational force between two objects depends on their masses and on the distance between them comes out naturally; these dependences are not merely postulated, as in Newton's theory. Indeed, physicists often define mass and energy as the stuff that gravity is connected to in Einstein's theory.

Incidentally, did you know that "Ashes, ashes, we all fall down" is from *Ring Around the Rosy,* a song of the plague?

GRAVITON SPECIFIES THEORY

Einstein's theory, being after all a theory of how the gravitational field behaves, specifies completely the properties of the graviton. Physicists have worked out that the graviton is massless, just like the photon, and spins twice as fast as the photon. They have also worked out how the graviton is emitted and absorbed by other particles. For the purpose of our discussion, it is not important to know in detail what the graviton's properties are. Suffice it to say that the graviton's properties are completely specified by the theory.

Now suppose we turn it around. Given the properties of the graviton, can you deduce that gravity exists?

Yes! Not only can you deduce its existence but also that it must be described by Einstein's theory. The properties of the graviton dictate the theory.

To see what this means, suppose you were marooned on an island

and to pass the time you decided to construct a theory of the world. While you know all the fundamental principles of physics, including the quantum principle, you have never heard of Einstein's theory of gravity. You scratch formula after formula on the sand. One day by chance you decide to include a massless particle that spins twice as fast as the photon and that can be emitted and absorbed by particles. You work out the consequences of having this particle in the world. Finally, a passing ship spots you and brings you back to civilization. After recuperating, you visit the local physics department and show them your theory. To your amazement, the theory you worked out is exactly Einstein's theory. (Legalistically speaking, your theory contains Einstein's theory; your theory may have more to it.)

In a sense, it is easy to get Einstein's theory. As long as a theorist includes a massless particle that has all the properties of the graviton, then bingo, he has Einstein's theory. This remarkable conclusion implies that it is very hard to alter the basic structure of Einstein's theory. The most you can do is to add to it.

IT MUST BE

The fabric of modern theories of physics is tightly woven. The reason is that the design and structure of these theories are mandated by deep, underlying symmetries. In the preceding chapter, I explained how the equivalence principle imposes a symmetry on the theory of gravity. Einstein's theory is constructed precisely to respect this symmetry. As a pale analogy, consider an architect told to design a building with a sixfold symmetry. Once she has decided on the design of one facet of the building, the other five facets are immediately fixed. So, too, if we are told that a massless particle with the properties of the graviton exists, the rest of the theory of gravity is fixed.

The analogy is imperfect in that the imposition of the sixfold symmetry still leaves the architect a great deal of freedom in the design. In contrast, the underlying symmetries of Nature are so intricate that they specify Nature. Symmetry dictates design. Once the symmetry underlying gravity was discerned, physics was literally forced to Einstein's theory.

Einstein's theory of gravity carries with it a sense of the inevitable. The notion that a particular theory is the only one possible was new to physics. For instance, Newton's pronouncement that the gravitational attraction decreases as the square of the distance between two bodies appears quite arbitrary from a purely logical point of view. Why doesn't the force decrease as the distance, or as the cube of the distance?

Newton would have regarded this question as unanswerable. He presents his law simply as a statement whose consequences accord with the real world. Altogether different, once Einstein understood the symmetry underlying gravity, the theory of gravity was fixed.

When I first encountered Einstein's theory of gravity, I marveled at how cleverly it is put together. With deeper understanding, I came to understand that it is essentially inevitable.

Abraham Pais, the leading biographer of Einstein, has aptly remarked that Einstein's theory of gravity has the full force of a Beethoven opus. The last movement of Beethoven's Opus 135 carries the motto, *"Musz es sein? Es musz sein."* (Must it be? It must be.)

Art in its perfection must be a necessity.

THE EXPANDING UNIVERSE

We follow the universe, the playground of gravity, from the big bang to the big chill.

Outward Bound

Anyone informed that the universe is expanding has a right to ask, "What's in it for me?"

—PETER DE VRIES

PAST THE STARS

In the early years of this century, it was decided that a great observatory was to be built. After eons of evolution, man was ready to look outward, past the stars. He wanted to stare at the universe itself. What he saw would change his conception of the universe forever. He saw that the universe was dynamic, almost alive, and that the universe had a beginning. He saw the expansion of the universe.

The drama of physics is populated with flesh-and-blood characters. For the enormous discovery of the expanding universe, the cast included a mule driver turned janitor, a heavyweight contender turned country lawyer, and a maniacal builder who drove his subordinates insane. We also have a Russian meteorologist, a Belgian *abbé,* and a great physicist who refused to see what his own theory told him.

THE MANIACAL BUILDER

We begin with the builder. George Ellery Hale made a telescope when he was a boy of thirteen, in 1881. Ever since then, he was obsessed with building larger and larger telescopes. Forcing himself relentlessly on the wealthy of the time, Hale managed to build, by the time he was twenty-nine, one of the world's largest telescopes, with a forty-inch lens. But soon he was dreaming of even bigger telescopes. Fortunately, Andrew Carnegie was willing to help.

After considering several sites, Hale chose Mount Wilson in the San Gabriel Mountains, overlooking Los Angeles. The location seemed ideal: The air was pristine, and the metropolis with its sparkling lights was still a developers' dream. In the winter of 1904, Hale and a carpenter, after fighting off a rattlesnake on the trail, reached the summit of Mount Wilson and set up a base. Later, a blizzard hit. Hale and an astronomer who had joined him were snowed in. Huddled around a fire, they spent the long winter nights planning the telescope.

The telescope was to have a curved sixty-inch mirror, almost inconceivable at the time. Few men had the skill to polish the monster glass disk to the exacting accuracy required. Hale found a man named George Ritchey with the necessary nerve and personality. I quote from an account given by D. O. Woodbury: "Ritchey was a lonely man who kept his own counsel and did not have many friends. . . . He shut himself into his optical shop and locked the door. Donning a surgical cap and gown, and sprinkling water on the floor to hold down the dust, he uncovered his mirror and began . . . with jeweler's rouge of the finest grade . . . the weary task of removing that last few millionths of an inch of glass." The glass could be polished only at a snail's pace, a few minutes at a time, lest the heat of friction twist the glass imperceptibly, but irrevocably, out of shape. One misstep and it would be over. The task took two years, and the stress on his mind was such that Ritchey was never the same again.

Meanwhile, Hale had to confront the wilderness of outer Los Angeles. The mountain was rugged, and equipment had to be hauled up by mules over an old Indian trail. The going was tough: "It was all right as long as the donkeys were not too fat; if they were, their stomachs rubbed the sides of the trail and were apt to start an avalanche, or else push the precious loads askew and send burro and all over the edge for a long roll back down the mountain." Later the track was widened to eight feet, and a specially rigged truck with a mule team hitched to it was introduced.

One of the builders, Jerry Dowd, described what it was like: "There was no rain on the last three miles of the trail and the dust and loose rocks were four inches deep—just like sugar. I would get up there with a load of beams or cement or something and stall every time. Then I would have to dig my way out. I took a man alongside of me on the seat with a shovel. He didn't do anything else but dig. He was worn out all the time. It took us ten days to make a round trip—fifteen miles up and fifteen back. . . . We hitched the mules to the front of the truck. They were pretty near human. They would just walk along, not doing a lick of work till they heard the engine begin to grind and labor. Then they'd take a strain on the chains and pull. When the engine calmed down they'd stop in a jiffy."

THE MANURE OF ST. GOBAIN

After much incredible travail, Hale finally saw the sixty-inch telescope installed in 1908.

What did Hale do then? He immediately decided he wanted a one-hundred-inch telescope. The man was obsessed, almost mad. Hale got a wealthy Los Angeles businessman named John Hooker to come up with the money. In the whole wide world, only one group of glassworkers could even imagine producing a glass disk this size: the royal glassblowers of the forest of St. Gobain in France; their group was founded in 1665 by Louis XIV. They had made the sixty-inch disk that Ritchey polished. I picture a group of wizened Frenchmen living amid the trees ever since the seventeenth century. *"C'est évidemment impossible, Monsieur Hale."* "But try!"

So they did. The five-ton slab was poured from three huge pots of molten glass and then buried in a manure pile for annealing. The manure was apparently the state of the art in this business. Only the French knew how to do it. But it was no use: The slab was full of trapped air bubbles. The elves of St. Gobain tried everything they knew and gave up. George Ritchey took a look and shook his head. Meanwhile, Hooker died. Even Hale lost heart and went back to his astronomy observations.

A year later, obsessed as always, Hale got the disk out of storage and convinced poor Ritchey to try again. This time it took him six years. There was no way that Ritchey could have tackled the job alone, and he reluctantly had to take on assistants. Now there was the additional strain that the forty-five-thousand-dollar glass slab could crack into a million pieces at any moment because of the air bubbles. Under the pressure, one of Ritchey's assistants went insane and had to be taken to an asylum.

Finally, in 1917, the Hooker telescope was ready. Ritchey refused to see it installed. He went out and sat on the edge of the cliff, staring into space. Then he left, never to set foot on Mount Wilson again. Hale himself paid dearly for his maniacal obsession. In 1910, he collapsed at an official function. For the rest of his life, he suffered from a type of brain congestion that produced severe pain and mental exhaustion. These people truly wanted to see the universe.

THE CURIOUS JANITOR

Among the mule drivers working for Hale was a dashing young fellow named Milton LaSalle Humason, known for his gamboling and for his way with ladies. Born in 1891 in Dodge Center, Minnesota, he had not managed to get beyond the eighth grade. In short order, Humason was courting the daughter of one of the engineers building the observatory. To be near her, he stayed on as a janitor after the observatory was built.

What Humason lacked in formal education, he made up with intelligence and curiosity. In between mopping floors, he found out what the astronomical observatory he helped build was all about. One night, in the middle of an observation, an astronomer felt ill, and Humason was ready. The janitor so astounded everyone with his skill with the telescope that he was officially promoted to assistant astronomer in 1919.

COUNTRY LAWYER TURNED STARGAZER

While Humason was being promoted, a major named Edwin Powell Hubble in the American occupation army in Germany was preparing to sail home. Two years older than Humason, Hubble grew up in Chicago, where his lawyer father was in insurance. In high school, the young Hubble excelled in both academics and athletics. He majored in mathematics and astronomy at the University of Chicago while establishing himself as a heavyweight boxer. He wasn't fooling around, either: A promoter wanted him to fight Jack Johnson, the then world champion. Funny, none of the math and astronomy majors I knew in college look even remotely like they could last a round or two with the world heavyweight champion. Times must have changed.

Hubble turned down the promoter, however, and decided instead to study mathematics at Oxford as a Rhodes scholar. But he soon switched to law. Boxing was not forgotten: He fought the French champion Georges Carpentier in an exhibition match. On the side, Hubble also won a number of track events. Sickening, isn't it, how some people have all the talents. Hubble was born a contender.

In 1913, armed with his Oxford law degree, Hubble opened a practice in Louisville, Kentucky. What the boxing world lost, the legal world almost won, but not quite. After a year of law, Hubble finally recognized his true love: "I chucked the law for astronomy, and I knew that even if I were second-rate or third-rate, it was astronomy that mattered."

He went back to the University of Chicago and in 1917 received a doctorate in astronomy. Already scintillating as an astronomer, he was immediately offered a position at the Mount Wilson Observatory. Just then, the United States entered World War I, and Hubble eagerly enlisted as a private in the infantry to risk life and limb in Europe. As gifted in war as in stargazing, Hubble rose quickly from private to major. (When World War II broke out, he again eagerly sought action but was persuaded to be the chief of ballistics at the Aberdeen Proving Ground in Maryland instead.)

And so finally in 1919, with Europe at peace, Edwin Hubble went to Mount Wilson to take up the position that had been reserved for him. The path of the mule-driver-turned-janitor-turned-astronomer and the path of the boxer-turned-lawyer-turned-astronomer crossed. The two men got along famously and entered into many fruitful collaborations.

Hubble's timing in starting his astronomy career was perfect. The one-hundred-inch telescope had just been completed on Mount Wilson, and with it Hubble was able to see farther and clearer than anyone before him.

Thus it came to pass that the man with all the talents had the world's largest telescope at his disposal.

CLOUDY PATCHES

For millennia, astronomers had noticed cloudy patches among the stars. (See Figure 4.1.) The general belief was that these so-called nebulae were relatively nearby and located among the stars. For a long time, astronomers thought that our galaxy comprised the entire universe and that the nebulae were perhaps gas clouds contained within the galaxy.

By 1920, however, some astronomers began to suspect that the nebulae lay extremely far away and were in fact each a galaxy like our own. If true, this would drastically revise our view of the universe. The universe would be much larger than previously thought, and our galaxy would be but one among many.

Using the new telescope, Hubble painstakingly measured the distance to the nebulae and established that indeed, the nebulae were galaxies. The announcement by Hubble in 1924 that there are galaxies far beyond our own marked the dawn of modern cosmology: We were able to look past the confines of our beloved Milky Way at the vast universe beyond.

Having found out how far the galaxies were from us, Hubble naturally wanted to know how fast they were moving.

4.1. *The Persian Abd-al-rahman Bin Omar Bin Muhammad Bin Sahl Abu'l-husain al-Sûfi al-Razi (A.D. 903–986), court astronomer to a powerful prince who ruled over most of modern-day Iran and Iraq, was the first to notice the Andromeda nebula, our friendly neighborhood galaxy. The woman represents Andromeda. The Arabic word* sahabi *coming out of the fish's mouth means "cloud-like."*

AN ORCHESTRA OF TRUMPETERS

I am always amazed that astronomers can actually measure the velocities of galaxies millions of light-years away just from the faint specks of light they manage to intercept. The remarkable properties of light make the feat possible.

Consider the familiar phenomenon in which the horn of a speeding train suddenly sounds lower as the train passes by us. The horn emits sound at a definite frequency. But as the train recedes from us, each successive crest of the sound wave has a slightly longer distance to travel to reach our ears. In a given duration of time, fewer crests reach us than would have been the case for a stationary train. We hear a lower-frequency sound. Conversely, as the train approaches us, its horn sounds higher-pitched.

Incidentally, the Dutchman Christoph Buys Ballot first tested this so-called Doppler effect in 1845 by having an orchestra of trumpeters perform in an open railroad car as the train sped through the countryside. Physicists had more style then—these days it would not be easy to get the National Science Foundation to pay for an orchestra of trumpeters. The demonstration I actually witnessed is less expensive but equally convincing. A radio humming at a definite frequency is placed inside a football. At the appropriate point in a lecture, the professor suddenly picks up the football and unleashes a long pass to a student seated at the back of the lecture hall. (The idea is to hit the dozing student on the head.) The class can clearly hear the hum becoming higher-pitched.

The Doppler effect applies to light waves as well as to sound waves. Since red light has a lower frequency than blue light, light from a receding source will appear redder. Light from a receding source is said to be *redshifted*. The greater the velocity of the source, the larger the shift. Conversely, light from an approaching source will appear bluer.

Wait a minute! you exclaim. The astronomer can measure the frequency of the light arriving at his telescope, but how does he know what the frequency of the light was when it left the star? If you don't know the intrinsic color of an object, how can you say it looks redder or bluer?

Good point. The Doppler effect would appear to be useless were it not for another fact. As sunlight passes from the inner core of the sun through the cooler outer layer, certain frequencies in the light are absorbed by gaseous atoms in the outer layer. Different types of atoms absorb different frequencies, as can be determined by laboratory experiments. Indeed, by analyzing sunlight to see what frequencies are missing, astronomers can ascertain what kinds of atoms are present in the sun. When astronomers analyze the light from a star, they find that the entire pattern of missing frequencies is shifted by a certain amount. Interpreting the shift as due to the Doppler effect, they can then determine how fast the star is moving away from us or toward us.

For a long time astronomers used this procedure to map out the motion of stars. Since the light from a galaxy is made up of the light from the individual stars contained in the galaxy, measurements of Doppler effects could also be used to determine the motion of remote galaxies.

THE FLIGHT OF THE NEBULAE

Already in the 1910s, the American astronomer Vesto Slipher had measured the redshift of a number of nebulae, as galaxies were then referred to, using a twenty-four-inch telescope at Lowell Observatory. At first, astronomers thought that the observed redshifts could be accounted for simply by saying that our solar system is moving away from these nebulae. Later, as this redshift was observed for nebulae in all directions of the sky, this interpretation became untenable. Furthermore, only redshifts were observed, never blueshifts (except for a few nearby galaxies). In 1918, the German astronomer Carl Wirtz proposed that for some reason, the nebulae were systematically moving away from us.

RAISIN BREAD AND OTHER ANALOGIES

Wake up in the morning and you see your neighbors' houses moving away from you. You would conclude that the earth is expanding and that some evil scientist must have thought of a way of literally blowing up the earth.

To establish that the earth as a whole is expanding, it is crucial to see how fast your next-door neighbor's house is moving compared to the house farther down the street. By the time the expanding earth has doubled in size, a house that used to be 50 yards away would be 100 yards away, a house that used to be 100 yards away would be 200 yards away,

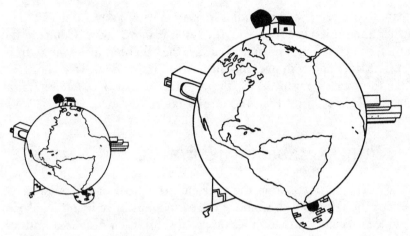

4.2. *Distances on an expanding earth.*

a house that used to be 150 yards away would be 300 yards away, and so on. (See Figure 4.2.) In that time, the house that used to be 50 yards away has moved 50 yards, the house 100 yards away has moved 100 yards, and so on. The farther the house, the faster it is moving away. More precisely, in a uniform expansion the receding velocity of a house should be proportional to its distance from us.

For another analogy, imagine yourself to be a raisin in a loaf of raisin bread as it gets baked. (See Figures 4.3a and 4.3b.) As the loaf

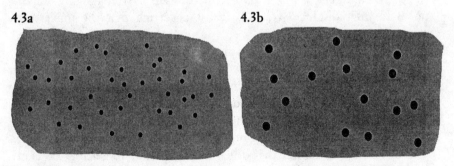

4.3a. *The drawing represents a cross section of raisin bread. The irregular boundary of the picture suggests that the raisin bread is infinite.* **4.3b.** *The raisin bread is expanded by a factor of 2.*

expands, you see the surrounding raisins moving away from you. The farther a raisin is away from you, the faster it moves away. Compare the motion of two raisins, one an inch away, the other ten inches away from you. The loaf doubles in size, and the nearby raisin is now two inches away while the faraway raisin is now twenty inches away. The faraway raisin moves ten times faster than the nearby raisin.

THE EXPANDING UNIVERSE

As we can see from the analogies, the proof of the pudding, or perhaps the raisin bread, is not in the eating, but in verifying if galaxies do indeed recede with velocities proportional to their distances from us. Only then can we be sure that we are actually seeing the universe expanding and not merely some local motion of galaxies in our immediate neighborhood. After all, even in a static universe, we expect galaxies to move around, in the same way that stars and planets move around under the influence of nearby massive objects.

Unfortunately for Slipher and Wirtz, they did not know reliably how far the nebulae were. Indeed, as I remarked, many astronomers at the time believed the nebulae to be some sort of strange objects within our Milky Way.

Hubble, however, had determined the distance to the galaxies and had established that they lie outside our galaxy. Thus he alone was able to put the velocity measurements and distance measurements together. In 1929 he announced that the velocities of galaxies were roughly proportional to their distances from us, thus suggesting that the universe is expanding.

Hubble's data, however, consisted of information about only eighteen galaxies. Furthermore, they were all fairly nearby in the cosmic scale of things. Serious doubts have been expressed as to whether Hubble could really reach his epoch-making conclusion with his rather scanty data. Perhaps he saw in his data what he wanted to believe—a common pitfall in science. His admirers, on the other hand, cite his boldness and insight. The cynics might suggest that other astronomers were getting too close for Hubble's comfort: Wirtz and his colleague K. Lundmark had already remarked in the 1920s that Slipher's redshifts tended to increase with the distances of the galaxies in question. Herein lies a fascinating question for the historian of science.

Hubble was keenly aware of the thinness of his data. To be more convincing, he had to measure the redshifts and distances of more remote galaxies. As I remarked above, galaxies move around in any case under the gravitational influence of nearby massive objects. This so-called peculiar motion—peculiar not in the sense of strange, but in the sense of characteristic of each galaxy—is superimposed on the motion due to the cosmic expansion. Since the cosmic motion due to the expanding universe increases with distance, the farther a galaxy is, the smaller its peculiar motion in comparison to its cosmic motion. In our expanding earth analogy, to make sure that the expansion is not limited to our neighborhood, we also have to look farther and farther out.

To probe the farthest reaches of the universe, Hubble enlisted the help of Milton Humason. As you can imagine, it is not easy to identify the missing frequencies in the light from a galaxy 100 million light-years away. Over the years, Humason the mule driver had become enormously skilled in analyzing these faint specks. Indeed, almost immediately, Humason was able to extend Hubble's data about four times farther. By 1931, Hubble and Humason had gone twenty times farther.

The mule driver pushed on steadily, and by 1936 he had measured the redshifts of galaxies some forty times more remote than those in Hubble's 1929 data. By then, the expanding universe was beyond doubt. The mule-driver-turned-astronomer played a crucial role in establishing the claim of the lawyer-turned-astronomer.

Humason's 1936 measurements had pushed Mount Wilson Observatory to its limit. The nightglow and the smog from Los Angeles were making observations increasingly difficult. Further progress had to wait until 1950, when the two-hundred-inch telescope on Mount Palomar was completed. By then, there was no need for mule drivers.

Guess who started the project to build the two-hundred-inch telescope?

NO MYSTERIOUS FORCE

A common misconception is that a mysterious force must be pushing the galaxies apart. To see the fallacy, think of a ball tossed upward. No force is pushing the ball and the earth apart. True, the tosser's hand exerts a force on the ball to impart to it a suitable amount of momentum, but once the ball leaves the tosser's hand, the only force acting on it is gravity, which tries to pull the ball back down. Whether gravity will succeed depends on how much momentum the ball was initially endowed with and on the mass of the earth. If the momentum is large enough, the ball will escape the gravitational clutch of the earth.

Similarly, some initial explosion, the "big bang," sent the galaxies flying away from each other, but the only force operating now between the galaxies is gravitational attraction.

Corresponding to the question of whether the tossed ball will escape from the earth is the question of whether the galaxies will keep on flying apart. To answer the former question, we need to compare the tendency to escape, as measured by the ball's velocity, with the tendency to capture, as measured by the earth's mass. Similarly, the answer to the cosmic question hinges on how fast the galaxies are receding from each other versus how much mass is contained in the universe.

If there is not that much mass in the universe, then the universe will expand forever. But if there is a lot of mass, then gravity will eventually arrest the expansion and bring the galaxies together again. The universe, after expanding to some maximum extent, will then contract. The galaxies will rush toward each other.

To follow the expansion of the universe, therefore, we need only a theory of gravity. There is no mysterious force of cosmic expansion.

THE UNIVERSE HAS NO CENTER

To understand the expanding universe more clearly, we must first understand a basic assumption of modern cosmology. Ever since Galileo and his intellectual followers braved the wrath of the Church and established that the sun does not revolve around the earth, the whole notion of anthropocentrism—that we humans occupy the center of the universe—has appeared ludicrous. Our sun turned out to be but a garden-variety star located near the edge of a garden-variety galaxy.

The scientific mind-set has swung completely over to the other extreme: We now feel that, over a suitably large scale, no region of the universe should be more special than any other. This general feeling is codified as the *cosmological principle*. All available evidence supports this principle. For example, the number of galaxies contained in a suitably large volume appears to be roughly the same anywhere in the universe.

This brings us to a popular misconception about the expanding universe: Galaxies are rushing outward from some central point. If so, then those galaxies near the central point would be more special than the galaxies farther out, contrary to the cosmological principle.

THE BIG BANG WAS EVERYWHERE

My friend who couldn't visualize curved three-dimensional space said, "Everything has a center! How can an object not have a center?"

"The problem is, when you think of a thing, you automatically picture a finite thing with edges. An infinite thing does not have a center. This is not some property of curved space, either. Just think of infinite Newtonian space extending out indefinitely. There is no center. Every point is like every other point."

"Yes, that's obvious. I get it, but only if the space is just sitting there. Not if there was a big bang. Boom! Boom!" She moves her arms to show how the big bang started. "In my mind's eye I can see that things start expanding outward from where the Boom! Boom! occurred. It sure looks like a central point to me, with the galaxies moving away from there, propelled by the explosion. I just don't get it."

"All right, picture the loaf of raisin bread."

"Yes, I can see it, sitting in the oven—"

"That's the problem; you are thinking of a finite loaf. As the loaf expands, the raisins near the bottom would not move much."

"Okay," my friend says, "I can picture the expanding loaf as somehow suspended in space, but I still tend to see the raisins as moving away from the raisin in the center. I know, I know what you are going to say—'Think infinite'—but my mind was made to think of finite loaves."

"Yes, think of a raisin-bread lover's idea of paradise, an infinite loaf filling all of space, so that there is no center and no edge. None of the raisins occupies a privileged position. As the loaf doubles in size, the distance between any two raisins doubles."

"Hmm—okay, but where was the big bang?"

"Good question. We can answer it using our raisin bread analogy. We picture an infinite and expanding raisin bread. Let's say it doubles in size every minute. Then a minute ago, the distance between any two raisins was half of what it is now. Two minutes ago, the distance between these same two raisins was . . . And three—"

"Aha, if you keep going, then eventually the raisins would be next to each other!"

"Exactly, and not only that, since those were *any* two raisins we were talking about, all the raisins were next to each other at that instant in time. That was the big bang!"

"So the big bang was where all the raisins were—"

"No; remember, think infinite, not finite! We have an infinite loaf with an infinite number of raisins. At the point of the big bang, the infinite number of raisins filled up all the space. It is a loaf of raisin bread without the bread!"

"I see, the big bang was everywhere. Hmm. But you don't have to have a big bang. When I pictured the loaf of raisin bread expanding, I never thought of it as expanding from when the raisins were all on top of each other. You physicists always take things to the extreme."

"But suppose we don't. I know you pictured a loaf with all the raisins comfortably separated from one other. Then you said 'Go!' in your mind and the loaf started expanding. You would have the universe starting with all the galaxies poised at some comfortable distance from their neighbors. Somebody waved the starting flag, and they all started moving away from each other. Surely you can't believe that."

"Well, yes, that would be kind of dumb." My friend looks sort of convinced. "But is the big bang merely a belief, a feeling that you can extrapolate all the way back?"

"No. Actually, there is strong evidence that the galaxies were once on top of each other. Why? Because we literally have fossils from that era, as we will see. Of course, we can never go all the way back to the very beginning."

Notice that curved space has nothing to do with the conclusion that the big bang was everywhere. Our entire conversation holds even if space is completely Newtonian and flat. These strange features of the expanding universe follow if the universe is infinite.

Some time later, my friend said excitedly, "I finally realized that the infinite is really different from the finite. A finite object, no matter how big, has a center, and an infinite number of raisins can fill up all of space even when they are all on top of each other!"

"You got it! I am reminded of the infinite hotel I once read about when I was a kid. A hotel with an infinite number of rooms is completely full: There is a guest in every room. But no sooner has the manager put up the No Vacancy sign when an infinite number of travelers shows up. What to do? The astute manager simply asks the occupant of room number 1 to move to room number 2, the occupant of room number 2 to move to room number 4, and so on. All the newcomers are then accommodated in the odd-numbered rooms. An infinite hotel is *not* just a very, very big finite hotel!"

FINITE BUT BOUNDLESS

Einstein's theory opens up another possibility. If space is curved inward like the surface of a sphere, then the universe could be finite without any edge. Our expanding-earth analogy illustrates the relevant point: The earth's surface is finite but has no edge. (Notice that we are talking about the earth's surface and not the solid ball that is earth.) Even though we see all these houses rushing away from ours, our house really does not occupy any special position. Our neighbor sees houses rushing away from his house. There is no center.

As we look back farther and farther in time, the earth shrank smaller and smaller, and the houses got closer and closer to each other. The analog of the big bang occurred when the earth's surface as a mathematical sphere shrank to a point. The big bang did not occur in the swamps of New Jersey any more than some other point on the earth's surface. It occurred when all the points on the earth's surface were one. In that sense, the big bang also occurred everywhere.

The analogy is not perfect because the earth's surface is two-dimensional. Without the aid of mathematics, it is hard to see how three-dimensional space can also be curved like the surface of a sphere. As I have said before, it is a matter of perspectives: A two-dimensional creature living in a curved surface would also have a difficult time seeing how his space could be curved.

BUSTING OUT

This brings us to yet another common misconception, that the universe is somehow expanding into a surrounding void. The most naive visualization of this erroneous view implies that the universe has an edge. A more sophisticated visualization pictures the expanding universe as finite but without edge. Consider the expanding-earth analogy. Wait, you might say, in the analogy the earth is expanding into a surrounding three-dimensional space, is it not? Yes, but to a two-dimensional creature living on the surface of the expanding earth, this fact is of no relevance. So, too, we might imagine our universe expanding into a surrounding higher-dimensional space, but if we cannot get outside the space we live in, then that description is not particularly useful.

On the other hand, what if we can? Sounds like science fiction, but no physical principle forbids the possibility that our universe may be contained in a larger space. As an analogy, picture a village in a long, narrow valley surrounded by steep mountains. To get out, one has to be energetic enough to climb over the mountains. You can easily speculate, as some physicists have, that we are also kept from getting outside the universe by an energy barrier. This speculation, while not serious physics, is not entirely idle, to the extent that the hypothesis can be tested experimentally. One suggestion is to smash two protons together with lots of energy and add up the energies of all the particles produced by the collision. If the energies added up do not equal the energy we put in, then we may want to say that some of the extremely energetic particles produced in the collision have busted out of the known universe. (Unfortunately, the test is not clear-cut, since the experimenter's instruments could simply have missed detecting some of the particles produced.)

TWO BRIGHT FELLOWS

Amusingly, two perfectly bright fellows, Newton and Einstein, each had a chance of predicting the expanding universe but blew it. The story of these missed opportunities illustrates the often overpowering influence the predominant philosophical outlook of an age can have on science.

Living three hundred years ago, Newton could hardly be expected to imagine anything but a static, unchanging universe. It was not until the 1930s that the cosmologist E. A. Milne pointed out that infinite Newtonian space uniformly filled with matter could describe the expanding universe, in a sense to be made precise below.

Incidentally, cosmologists customarily describe the universe as uniformly filled with matter in the same way that we can think of a gas as of uniform density, even though it is made of molecules whizzing about with enormous amounts of space between them. Looked at on a cosmological scale, even galaxies appear as points, and there are so many of them that the universe may be described as a uniform gas of galaxies.

Newton had in fact contemplated space uniformly filled with matter as an idealized model of the universe, but he and his contemporaries were puzzled over what would happen to such a universe. They recognized clearly enough that a finite ball of matter would simply contract toward its center under the force of gravity. (See Figure 4.4a.) But infinite Newtonian space uniformly filled with matter threw them for a loop. There is no center to which the matter can contract! It would appear that the speck of matter marked S in Figure 4.4b is tugged equally in all directions and hence would not move anywhere.

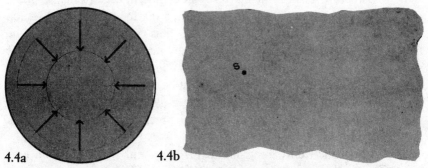

4.4a   4.4b

4.4a. *There is a big difference between a finite ball of matter however big it may be, and an infinite distribution of matter filling up all of space. A finite ball of matter would collapse toward its center. In the infinite distribution of matter represented in* **4.4b** *as a distribution with an irregular edge, which way would the speck of matter marked S move?*

CENTER, CENTER, EVERYWHERE

God is a circle whose center is everywhere and whose circumference is nowhere.

EMPEDOCLES, fifth century B.C.

The fabric of the world has its center everywhere and its circumference nowhere.

CARDINAL NICHOLAS OF CUSA, fifteenth century

Milne pointed out that since there is no center, the center can be everywhere!!! Sounds like Zen, doesn't it? But it is actually perfectly sensible physics.

To understand this remarkable piece of physics, we have to remind ourselves of a theorem about gravity discussed in the Prologue, a theorem discovered by none other than Newton: The gravitational force exerted by a spherical shell on an object inside it is zilch. Remember how we applied the theorem to describe the gravitational force we would feel as we ride the gravity express to the other side of the earth?

Look at the speck of matter S in Figure 4.5. Consider a sphere centered at some other point C—it doesn't matter which point—such that S lies on the surface of that sphere. This construction divides the matter in the universe up into two parts, the matter contained inside the sphere and the matter outside the sphere. The matter outside the sphere can now be divided into an infinite number of spherical shells. But we have just learned from Newton that the pulls exerted on the speck of matter by these spherical shells all add up to precisely zero. Thus the speck S is pulled toward the center C.

The really weird thing about this argument is that you can pick the center C to be anywhere. The speck S is pulled toward anywhere! This statement would not have made any sense had we not gone through the raisin bread example. The center is everywhere also in a contracting *infinite* loaf: Every raisin is moving toward every other raisin.

When we discussed the expanding loaf of raisin bread, we found it convenient to look back and ask where the raisins were some time ago. In effect, we thought of the loaf as contracting. Physicists also often find it useful to picture a contracting universe, as we just did. We can then simply reverse the motion of the matter in our picture to obtain the expanding universe. (To think of the contracting universe is to think of the falling apple, while to think of the expanding universe is to think of the apple tossed upward. The latter is a trifle more confusing.)

The discussion of this section shows how a Newtonian universe uniformly filled with matter would contract under the force of gravity.

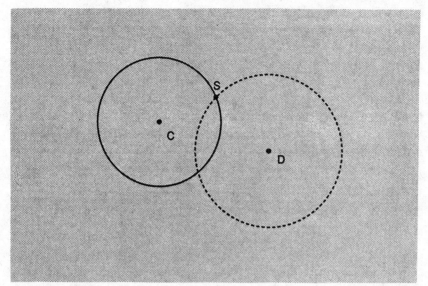

4.5. *"The center is everywhere": We are interested in the motion of the speck of matter marked S in a Newtonian universe uniformly filled with matter. According to the analysis in the text, we can pick a point C ("the center") and draw a sphere centered at C and passing through S (solid line). The speck of matter at S feels the gravitational pull of the matter contained within this sphere. But we could repeat the construction with any other point, say the point D, and draw a sphere centered at D and passing through C (dotted line).*

Every speck of matter would move toward every other speck of matter. Every galaxy would move toward every other galaxy. When reversed in time, this was precisely what Hubble and Humason saw.

Thus Newton could have discovered the expanding universe as we now understand it.

A CONTRACTING UNIVERSE

As the universe expands, how does the expansion rate change? This question can sometimes confuse even a professional physicist. "Hmm, as the universe expands, the matter in the universe gets thinned out, and hence the effectiveness of gravity should diminish," he might say. "Since gravity was acting to brake the expansion, therefore the expansion should speed up." The fallacy of this argument is revealed clearly by the trick of picturing a contracting universe rather than an expanding universe. As a universe contracts, the matter in the universe gets denser and denser; hence gravity is more and more effective in pulling the matter together. Thus the universe contracts faster and faster. Now simply reverse the motion of the matter, and we see that an expanding universe expands slower and slower.

The result that our universe was expanding much faster in the past will be important to us later.

THE WORST BLUNDER

What irony! Newton himself had invented the theorem needed for Milne's argument. But as I have already said, give the fellow a break, he lived so long ago. What is Einstein's excuse?

Einstein's excuse is really no better than Newton's. Nothing in the years between Newton and Einstein had cast serious doubt on the belief in a static universe. When Einstein decided in 1917 to use his equation of gravity to describe the universe, he didn't bother to solve his equation. He simply checked to see if his equation conformed to his own philosophical prejudice. Well, a static, unchanging universe is not a solution of the equation. Upset, he proceeded to change his own equation. In effect, he arbitrarily invented a new force, which he carefully balanced against gravity to prevent the universe from expanding or contracting. Later, with the universe's expansion established, Einstein was supposed to have said that his mutilation of his own theory was the worst blunder of his life.

While Newton simply failed to see the relevance of his own theorem for cosmology, Einstein actively imposed his belief on his own equation.

ABOMINABLE PHYSICS

For hundreds of years, the physics of gravity had been crying out to anyone who could listen that the universe cannot be static. In a sense, that the universe cannot be static is obvious: Someone would have to hold the galaxies in place.

Finally, in 1922, the Russian metereologist Aleksandr Aleksandrovich Friedmann (1888–1925) solved Einstein's equation and discovered a solution describing an expanding universe. Einstein thought at first that Friedmann's work must be wrong, but he was unable to find an error.

Unaware of Friedmann's work, the Belgian *abbé* Georges Lemaître (1894–1966) also discovered, in 1925, that Einstein's theory implies an expanding universe.

The *abbé* is the last character to come onstage in our drama of the expanding universe uncovered. When the guns of August boomed, the young Lemaître, having just become a mining engineer, joined a corps of volunteers and engaged in bloody house-to-house fighting. Later he became attached to an artillery regiment of the Belgian army but got himself removed from officers' training school for pointing out an error in the instructor's solution to a ballistics problem. I wonder how close the paths of Lemaître and of Hubble came to each other during the war. After the war, Lemaître switched careers and obtained a doctorate in mathematics in 1920, whereupon he decided to enter a seminary. He was ordained a priest in 1923. A month later, he was abroad studying astronomy and soon made the great discovery that Einstein had refused to see.

Incidentally, a curious footnote: Lemaître's father owned a glassworks. For years he experimented with new processes of dealing with molten glass until he blew up the factory in an accident, sending the family into bankruptcy. (Perhaps he was competing with the elves of the Forest of St. Gobain?) When my friend heard of this, she said with a laugh, "The grandfather of the big bang blowing himself up? I love it!"

At a conference in Brussels in 1927, Lemaître tried to tell Einstein about his discovery, but Einstein dismissed him curtly in French: *"Vos calculs sont corrects, mais votre physique est abominable."* ("Your calculations are correct, but your physics is abominable.")

The idea that the universe may expand was not taken seriously until Hubble's discovery in 1929. Unfortunately, Friedmann had died in 1925, a dramatic death we will come back to later.

Darkness at Night

THE WINGS OF NIGHT

The day is done, and the darkness
Falls from the wings of Night.

—H. W. Longfellow

Deep into that darkness peering,
Long I stood there wondering, fearing.

—E. A. Poe

Night after night, the darkness came. Strange noises filled the jungle, and our ancestors huddled closer together. For eons, the darkness of night frightened and mystified us, but nowadays, nobody thinks the darkness puzzling. Every person in the street knows, presumably, the explanation taught to schoolboys: The earth rotates on its axis as it moves around the sun, and every twenty-four hours night falls on that half of the world facing away from the sun.

Remarkably, this explanation is seriously incomplete. That it is incomplete was realized more than two hundred years ago, but for some reason continues to be kept secret from the proverbial schoolboy. The correct explanation, or at least what we now believe to be the standard explanation, was not completely clarified until the 1960s. Astonishingly enough, the correct explanation of this apparently simple phenomenon, that it gets dark at night, tells us a great deal about the universe.

Where is the flaw in the standard explanation? It lies in the implicit assumption that the sun is our only source of light. What about the stars?

But the stars are very far away, you say, and hence very dim. We all know, for instance, that during the day the sun completely overwhelms the stars and they become invisible. That's true, but aren't there zillions and zillions of stars? Each star sends us only a tiny amount of light compared to the sun, but couldn't the light from all the stars in the universe add up to a big effect?

THE ENLIGHTENMENT MIND

Familiar things happen, and mankind does not bother about them. It requires a very unusual mind to undertake the analysis of the obvious.

—ALFRED NORTH WHITEHEAD

To see whether the light from all the stars is truly significant, we have to do a simple calculation, first done by the Swiss Jean Philippe de Cheseaux in 1744 and independently by the German Heinrich Olbers in 1826.

Let us then look over de Cheseaux's shoulders as he calculates. The year is 1744; the place, Lausanne; and the king of France, Louis XIV. Voltaire (1694–1778) is making fun of everything held sacred, and the Enlightenment is in full swing. The informed no longer believe that our solar system occupies a privileged position. The celestial spheres with their music have been replaced by the infinite space and time of Newton (1642–1727) and Blaise Pascal (1623–62), filled with a uniform distribution of stars, each not unlike our own. (Incidentally, whether the universe is infinite was once a heated topic indeed. The monk Giordano Bruno was burned at the stake in 1600 for suggesting an infinite universe.)

We know well the laws of optics, thanks to such men as Newton, Robert Hooke (1635–1703), Christian Huygens (1629–95), and Francesco Grimaldi (1618–63). As they have shown, it is a simple matter of geometry to determine how the brightness of a light source decreases with increasing distance. Consider a star, sending its light out in all directions

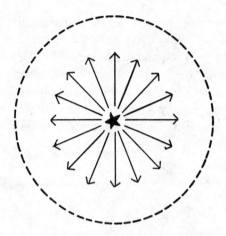

5.1. *Light streaming from a star crosses a mathematical sphere drawn around the star. (This is a two-dimensional drawing of a three-dimensional situation. The dotted circle is a cross section of the sphere.) Clearly, each unit of area on the surface of the sphere catches an equal amount of light.*

(Figure 5.1). To determine the fraction of light actually caught by an observer some distance away, we draw a sphere centered on the star and with a radius equal to the distance between the star and the observer.

Each unit of area on the surface of the sphere catches an equal amount of light. Therefore, the fraction of starlight caught by the observer is equal to the area of his eyes divided by the surface area of the sphere. As any photographer knows, the larger the aperture, the more light that enters the camera.

How much less light would be caught by an observer with the same size eyes but located farther away?

Well, we have just established that the fraction of light caught is equal to the area of the eyes divided by the surface area of a sphere with its radius equal to the distance between the star and the observer. Thus, the farther away the observer, the larger the relevant sphere and hence the smaller the fraction of light caught by him. This is illustrated in Figure 5.2, with an observer twice as far away from the star as another observer.

As our proverbial schoolboy knows, a sphere twice as big as another has four times as much surface area. The surface area of a sphere is proportional to the square of its radius. Thus an observer twice as far away as another from a star will perceive the star to be one fourth less bright. Similarly, an observer thrice as far away will perceive the star to be one ninth as bright. And so on.

Now that we have established how much dimmer distant stars are, we turn to counting them. To do so, imagine an enormous spherical shell

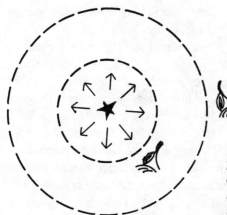

5.2. *An observer twice as far away from a star as another observer catches only one quarter as much light. Just as in Figure 5.1, you have to imagine the spheres.*

centered on us. (See Figure 5.3.) The thickness of the shell is to be small compared to the scale of the universe but large compared to a star, so that the shell contains a multitude of stars. As we look into the sky, we catch a small fraction of the light emitted by each of the stars within our field of view. Now imagine a second shell with a radius twice that of the first. According to the preceding, a star contained in this second shell will look only one fourth as bright as a similar star contained in the first shell.

But, and here is the punchline of the paradox: There are also

5.3. *Counting stars: Imagine two thin spherical shells, one with a radius twice as large as the other, as described in the text. In order to show one spherical shell nested inside the other, the artist has cut away half of each shell. If the two shells have the same thickness, the larger shell would have four times as much volume as the smaller shell. If the universe is uniformly filled with stars (not shown here), then the larger shell would contain four times more stars in the thickness of its shell than the smaller shell.*

exactly four times as many as stars within our field of view in the second shell as in the first shell! The *four* cancels the *one fourth*. To us, the sum-total brightness of the stars in the second shell is exactly equal to the sum-total brightness of the stars in the first shell.

We can repeat the argument for any one of the infinite number of shells drawn around us. The sum-total brightness of the stars within our view in any shell is the same regardless of the shell, be it close at hand or far away.

Now we see the paradox. We can envisage infinite Newtonian space divided up into an infinite number of shells, much like an infinite onion with us at the center. Since each shell contributes the same amount of brightness to the sky, the cumulative effect of all the shells is an infinitely bright sky.

And yet the darkness comes night after night.

LUCIDITY BURIED

This is essentially the argument as originally given by de Cheseaux, an argument so lucid that it can hardly be improved on after two centuries: Simply, the dimness of the distant stars is exactly compensated for by their number. Unfortunately for him, he buried his remarks in an appendix to a book on an entirely different topic, and they remained unknown to the scientific community until 1960.

Some eighty years after de Cheseaux, Heinrich Olbers rediscovered the same paradox. Olbers was sixty-eight when it occurred to him that he didn't know why it got dark at night.

THE GOLDEN WALLS

Look down into the abysmal distances!—attempt to force the gaze down the multitudinous vistas of the stars, as we sweep slowly through them thus—and thus—and thus! Even the spiritual vision, is it not at all points arrested by the continuous golden walls of the universe?

—E. A. POE

Following de Cheseaux and Olbers, we took the stars to be geometrical points for simplicity. To take into account the finite size of stars and the fact that a star can block the light emitted by stars behind it, think of the lines of sight from our eyes to a star. (See Figure 5.4.) Along each line of sight, a bit of light travels from the star to our eyes. The apparent brightness of the star is proportional to the fraction of our lines of sight that terminate on it: the farther it is, the smaller that fraction. But in an infinite universe, every line of sight terminates on a star. And hence we conclude that the entire sky should be as bright as the surface of the average star, a milder but equally shocking conclusion. The heavens should be brightly lit all the time—great for our nightlife, if the word *night* still means anything, but probably terrible for our health. It would be unbearably hot, with the temperature the same as the surface temperature of the average star.

When theoretical expectation disagrees so violently with an observational fact we reconfirm night after night, clearly some of our assumptions must be incorrect. But which ones? We can immediately check that the minor assumptions are not responsible for the paradox. For

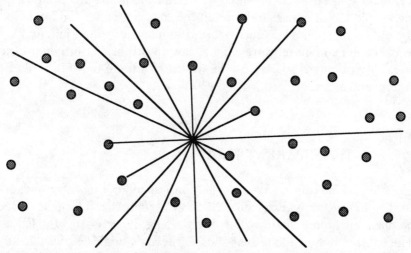

5.4. *In an infinite universe populated uniformly with stars (represented here by dots), the lines of sight from any point will all eventually terminate on a star. Nine of the lines of sight drawn did not terminate on a star within the figure, but if continued indefinitely they would eventually terminate on a star.*

instance, we assumed the stars to be intrinsically equally bright. In fact, some stars are significantly brighter than others. But obviously the argument still goes through, provided that statistically there are in any given region bright and dim stars in equal measure.

ESCAPE ROUTES

How can we escape from this paradox?

De Cheseaux himself suggested two ways out. The first supposes that the universe is finite, a suggestion de Cheseaux rejected immediately because to keep the night sky dark, the universe would have to be incomparably smaller than what he thought reasonable. A finite Newtonian universe is unacceptable to the Enlightenment mind because it implies an edge to the material universe. Already, the Roman poet-philosopher Lucretius had remarked: "The universe is not bounded in any direction. If it were, it would necessarily have a limit somewhere, but clearly a thing cannot have a limit unless there is something outside to limit it." Later, Cardinal Nicholas of Cusa, whom we heard from in the preceding chapter, argued for an infinite universe on theological grounds. To him, God is clearly infinite, and it is surely absurd to imagine that an infinite God would create a finite universe. (It was not until Einstein introduced the possibility of curved space that the notion of a finite universe could be entertained again.)

A TRANSPARENT FLUID

De Cheseaux's second suggestion was more interesting and, according to him, fairly reasonable. "It demands only that interstellar space be filled with some fluid capable of absorbing, however slightly, light." The contribution of the distant stars to the sky's brightness would then be decreased. He went on to state that a fluid 3×10^{17} times more transparent or diffuse than water would do the job.

Why this escape route is also blocked is somewhat more subtle. According to the laws of thermodynamics—a subject not well understood in the eighteenth century—an interstellar medium capable of absorbing starlight would eventually become so hot that it would start emitting light as fast as it absorbs the light.

THE BEGINNING OF TIME

We have considered and rejected the possibility of a finite-size universe. What about a universe finite in time? Implicit in de Cheseaux's argument is the assumption of an eternal universe that has existed forever. But if the stars started shining only a while ago, then the paradox can be resolved. Suppose, for instance, that de Cheseaux would accept Archbishop James Ussher's calculation that the universe was created in the year 4004 B.C. Then we escape the paradox easily, since the light from stars farther away than six thousand light-years would not have reached us yet.

While we chuckle at Ussher, I may mention that Newton himself came out with a figure of 3988 B.C. for the creation of the universe. Scientists, like everybody else, are products of their times.

Primitive people quite naturally speculated on a beginning for the world, and every culture has its own creation myth. But scientists from the Enlightenment to the time of Einstein were reluctant to think of the universe originating in a special event. They could not and did not take the likes of Archbishop Ussher seriously. They thought of the universe as static, existing in the unceasing flow of Newtonian time, with neither beginning nor end. The notion universally entertained by primitive people that the world had a beginning was brought back only with the discovery of the expanding universe.

Distant galaxies are rushing away from us. Run the film backward, and they rush toward us. At some point in the past they must have been on top of us.

Hubble and Humason discovered that galaxies are receding at velocities directly proportional to their distances from us. The proportionality factor is now thought to be about 30 kilometers per second per million light-years. Wait a minute—too many *pers* here. What does this number mean? Simply, a galaxy a million light-years away from us is receding at 30 kilometers per second, a galaxy two million light-years away from us is receding at 60 kilometers per second, and so on.

Consider the galaxy that is now a million light-years away from us. How much time has passed since it was on top of us?

As a first try, assume that the expansion rate has always been the same. Well, the answer is clearly the time it takes an object traveling at 30 kilometers per second to cover a million light-years. Divide a million light-years by 30 kilometers per second. Oh, no, we have to convert units; that always drove me crazy in school. Let's look up in a suitable handbook that 1 light-year $= 10^{13}$ kilometers and 1 year $= 3 \times 10^7$ seconds. Good. Hmm, let's see, a million light-years, that is 10^{13} million kilometers, divide that by 30 kilometers per second and we get about 3×10^{11} million seconds. Convert that to years. Divide 3×10^{11} million seconds by 3×10^7 seconds in the year, and we get 10^4 million, that is, 10,000 million, or in American English, 10 billion. Whew, did we get that right? Believe me, in my theoretical work I don't have to do arithmetic that often.

So, ten billion years ago, the galaxy that is now a million light-years away was right on top of us. In fact, every galaxy in the universe was right on top of us! Think about it. The galaxy that is two million light-years away moves twice as fast as the galaxy one million light-years away, the galaxy three million light-years away moves thrice as fast, and so on.

Galaxies are not special in this context. As in the expanding raisin loaf, everything moves away from everything else in the expanding universe. A Buick a million light-years away from us would also be receding at 30 kilometers per second. Indeed, ten billion years ago, the distance between any two points in the universe was exactly zero, a situation mathematicians describe as a singularity. This singular event was the big bang.

The expanding universe implies a beginning.

Therefore the darkness comes, night after night, even if the universe is infinite in spatial extent. We have yet to see the light from those stars more than ten billion light-years away.

The good archbishop had the right idea; he was off by a factor of only 1 million or so.

THE AGE OF THE UNIVERSE

Let us now examine the assumption made above that the expansion rate has always been the same. In fact, we expect the expansion to be faster in the past because, as explained earlier, gravity has been steadily slowing down the expansion. Therefore, our value of ten billion years for the age of the universe is actually an overestimate. A more accurate estimate requires a calculation using Einstein's equation of gravity. For our purposes here, ten billion years is close enough.

Historically, there was a foul-up. Hubble and Humason published a value for the expansion rate about ten times larger than the value now believed. Thus, the universe came out to be only one billion years old. But already at that time geologists were pretty sure, because of studies of natural radioactivity, that the earth was about five billion years old. The discrepancy caused much bewilderment before astronomers finally straightened out their date.

Incidentally, I should also mention that some astronomers believe that the expansion rate is only 15 kilometers per second per million light-years. If so, then the universe would already be celebrating its twenty-billionth birthday now.

A HIDDEN GOD

When I first read about the paradox of the night sky, I was flabbergasted. Why is the sky dark at night then? When I learned of the resolution, I was astonished. Every time we look up into the bowl of night, we see the beginning.

Had there been someone in the human race as smart as de Cheseaux but unfettered by philosophical prejudices, he or she could have predicted, on the basis of a common every-night observation and before Hubble and Humason, that the universe had not always been.

To appreciate the full significance of the night sky paradox, we must place ourselves in the proper historical perspective and understand how alluring the notion of an unchanging, eternal universe was for thinkers until recent times. That the universe began with a big bang has been so well publicized that it now appears as a joke in popular movies. It is the framework into which we fit our cosmological ideas, but it was not always so.

Throughout the 1930s and 1940s, physicists ridiculed the idea that the universe had a beginning. The foul-up of Hubble and Humason certainly did not help. The subject was inevitably charged with theological overtones.

Since Friedmann had died, Lemaître became the principal champion of what he referred to rather quaintly as the Primeval Atom Hypothesis. It had been said that physicists regarded the theory with even more suspicion than otherwise because Lemaître was a priest. Some saw the primeval atom as but a thinly disguised deity.

Lemaître felt defensive and wanted to have his work judged solely as physics. In this regard he was embarrassed by the interest of Pope Pius XII in physical cosmology. At a congress at the Vatican in 1951, the pope, after giving a detailed exposition on physics and cosmology, declared that the big bang "could be made the antecedent of the scholastic syllogism concluding to the Catholic concept of creation."

Lemaître prevailed upon his friend Father Daniel O'Connell, a Jesuit director of the Specola Vaticana, to exercise his influence on the pope, particularly since the pope was scheduled to address the International Astronomical Union in 1952. This time the pope applauded the progress made in observational astronomy but refrained from mentioning the big bang. Father O'Connell apparently did his work.

Lemaître was relieved, but he felt obligated to explain his position to his fellow physicists. He wrote: "As far as I can see, such a theory remains entirely outside any metaphysical or religious question. It leaves the materialist free to deny any transcendental Being. He may keep, for the bottom of space-time, the same attitude of mind he has been able to adopt for events occurring in nonsingular places of space-time. For the believer, it removes any attempt to familiarity with God, as were Laplace's chiquenaude or Jeans's finger. It is consonant with Isaiah's speaking of the Hidden God, hidden even in the beginning of creation."

THE LURE OF THE UNCHANGING

It is within such a charged atmosphere that we can understand the overpowering desire to have an eternal universe, to avoid any talk of a beginning. This desire drove Hermann Bondi and Thomas Gold in 1948 and Fred Hoyle somewhat later to invent the steady state theory. For the universe to maintain an unchanging face as it expands, matter has to pop out of literally nowhere and form into galaxies to fill the space opened up by the receding galaxies. Otherwise, the number of galaxies per unit volume would not stay constant. Thus the proponents of the steady state theory had to mutilate arbitrarily the known laws of physics to allow for the continuous creation of matter from nothing. Creation was replaced by an ongoing creation. Many physicists found this rather distasteful. Nevertheless, the steady state theory was in vogue for quite a while.

A major astronomical discovery in 1965 finally ruled out the steady state theory, as we will see in Chapter 6.

Historically, the night sky paradox came back into the collective consciousness of physicists because of the steady state theory. It posed as serious a difficulty for steady state cosmology as it would for any cosmology with an eternal universe. Much of the effort of the steady state theorists was devoted to resolving the paradox.

The lure of the unchanging remained strong. In 1967 Dennis Sciama wrote wistfully: "For me the loss of the steady state theory has been a cause of great sadness. The steady state theory has a sweep and beauty that for some unaccountable reason the architect of the universe appears to have overlooked. The universe is in fact a botched job, but I suppose we shall have to make the best of it." It's a matter of opinion.

THE BLUE MUNDANE SHELL

There, now we understand why the sky is dark at night. But why, then, is the sky bright during the day?

The sun is shining, that's why, our proverbial schoolboy answers. That's in fact correct, but somewhat incomplete. Should the earth not have an atmosphere, the daytime sky would look quite different. We would see the sun as a ball of fire suspended against a black sky. In almost every direction in which we looked, if our lines of sight did not end on the sun, the sky would be dark.

As sunlight streams through the atmosphere, it scatters off the air molecules. Much as a lampshade moderates the harshness of the bare bulb, the atmosphere, that "blue Mundane shell that separates us from Eternity," in the words of William Blake, acts to bathe us in a soft blue glow. (The sky is blue because the light waves corresponding to blue are scattered more than those corresponding to red and yellow.)

HOW FAR THE GOLDEN WALLS

Actually, the night sky paradox was already solved in 1901 by the physicist Lord Kelvin, before the discovery of cosmic expansion. He pointed out that stars do not burn forever. Recall the argument that any line of sight from our eyes traveling outward into the universe will eventually land on the surface of a star. (See Figure 5.4.) The distance our line of sight has to "travel" before it lands on a star obviously depends on its direction. Consider the average distance, which I will call the distance to the golden wall, in honor of Edgar Allan Poe. Clearly, said Kelvin, the paradox hinges on comparing the time it takes for light to traverse the distance to the golden wall versus the lifetimes of the stars. We will see why in a moment.

Knowing the sizes of the stars and how densely they are distributed in the universe, we can easily calculate the distance to the golden wall. Clearly, the more sparsely the stars are distributed, the larger this distance. Allowing for the fact that stars are gathered into galaxies, we find 10^{23} light-years. In contrast, we now know that a typical star like the sun lives for only 10^{13} years.

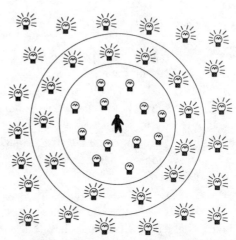

5.5. *Even in a static Newtonian universe uniformly filled with stars, the finite lifetime of the stars, here represented as light bulbs, would limit the brightness of the sky. Let all the light bulbs in the universe be turned on at once. Suppose the bulbs burn out after one year. The figure shows what you would perceive three years after the light bulbs were turned on. You could see that all the bulbs within a sphere of radius two light-years (the smaller sphere in the figure) had been burned out. Meanwhile, light from the bulbs that are three light-years away (the bigger sphere) are just beginning to reach you. At any given time, you would see only the light from light bulbs within a spherical shell with a thickness of one light-year.*

Imagine yourself in this hypothetical eternal static universe in which the stars, like so many light bulbs strung throughout the firmament, were all turned on at once. Would the dazzling brilliance of this infinitude of stars instantly blind you and convert you to religion?

Not at all! exclaimed Kelvin.

To blind you, *all* the stars within a volume of 10^{23} years have to hit you with their light. But after standing there looking at the sky for 10^{13} years you will begin to receive the light only from those stars 10^{13} light-years away. Now, if the stars burn forever, you will eventually be in trouble. Fortunately, the stars are now burning out. You stop seeing the nearby stars. Even though the distant stars are also dead, you continue to see them by the light that left them before their death. But if a star is too distant, its light has yet to reach you. Clearly, at any given time you perceive only the light from the stars within a spherical shell 10^{13} light-years thick. (See Figure 5.5.)

Since we now know that the universe is itself only 10^{10} years old, the dark night sky does not present, a fortiori, a paradox. There is no need to invoke the finite lifetimes of the stars. Furthermore, no one thinks now of an eternal, static universe in which all the stars are turned on at once. (Stars are created continually from interstellar gas.) Indeed, Kelvin's neat little argument had been largely forgotten and was unearthed only recently by the astronomer Edward Harrison.

Kelvin's resolution underscores an important distinction between mathematics and physics. The statement that any line of sight from our eyes will eventually land on a star, given that stars are of finite size, and that stars are uniformly distributed in space, is an indisputable mathematical truth. But it took a physicist, and one as close to actual numbers as Lord Kelvin, to realize that the relevant distance comes out to be outlandishly large. The mathematician is concerned only that it is "eventually" and not "never." The physicist, being told "eventually," wants to know "how soon."

THE POET'S SPIRITUAL VISION

When I heard the learn'd astronomer,
When the proofs, the figures, were ranged
    in columns before me,

. . .

How soon unaccountable I became tired and sick,
Till rising and gliding out I wander'd off by myself,
In the mystical moist night-air, and from time to time,
Look'd up in perfect silence at the stars.

—WALT WHITMAN

Intriguingly, Edward Harrison, in his detailed historical study of
the night sky paradox, suggested that Edgar Allan Poe was probably the
first to resolve it correctly. In 1848, three years after he exulted over how
even spiritual vision is arrested by the golden walls of the universe, Poe
wrote that the only way out is "by supposing the distance of the invisible
background so immense that no ray from it has yet been able to reach us
at all." Poe's qualitative resolution anticipated Kelvin's quantitative resolution by half a century.

In a sense, then, the paradox of the dark night sky is resolved
simply because there are so *few* stars. On a recent summer evening, I went
up into the mountains high above Santa Barbara with my family to have
dinner. The restaurant, converted from an old stagecoach stop, was isolated from civilization. As we dined, the earth gradually turned away from
its sun. When we emerged from the restaurant, I was suddenly impressed.
I had forgotten how many stars there are, yet how impenetrably dark the
cloudless night sky can look.

Yes, the stars are beyond counting, but there are big patches of
darkness between them. Were I to fly straight into that darkness, I would
have to go for, on the average, 10^{23} years, during which time our universe
could have relived its life 10^{13} times over, before I would plunge into the
fire of a star.

Too few stars, too few indeed, compared to the frightening and
incomprehensible vastness of space. I recalled the words of Pascal: "The
eternal silence of infinite space frightens me."

From the Big Chill to the Big Bang

THE BIG CHILL

Poor Aleksandr Aleksandrovich Friedmann! In his picture, he looked so sad and so much the image of the nerd mathematician. (6.1)

But looks may be deceiving: Friedmann, the great Russian cosmologist who saw the expanding universe in Einstein's equation, had a streak of the adventurer in him. He was smitten with the notion that he could predict the weather mathematically were he able to measure the detailed physical conditions within a large volume of the atmosphere. And so he went up in a balloon. He caught a severe chill, then pneumonia, and died. He was thirty-seven.

6.1. *Aleksandr Friedmann, a Russian hot-air balloonist.*

EXPANDING AIR COOLS

I tell you the story of poor Aleksandr Friedmann because it actually has something to do with the expanding universe. As a given amount of air moves up into higher altitudes, it expands because there is less atmospheric pressure squeezing on it. Atmospheric pressure is, after all, due to nothing more than the weight of the air above us in the earth's atmosphere pressing down. As air expands, it cools.

Why does an expanding gas cool? In the nineteenth century, physicists identified the hitherto mysterious phenomenon of heat as merely the manifestation of the agitated movement of molecules. The faster the molecules in a gas zoom about, the hotter the gas. (See Figure 6.2b.) The temperature of a gas is simply the *average* energy of the molecules. To be sure, at any given temperature, some molecules are moving faster than the average, others slower than the average. Now allow the gas to expand. As it expands, its molecules lose energy as they push the walls of the container outward. In the whimsical analogy pictured in Figure 6.2b, suppose the walls of the classroom are mounted on wheels. Every time a kid bounces off a wall, the wall moves and gains energy.

6.2a

6.2b

A whimsical representation of a gas at two different temperatures, with the gas molecules replaced by school children. **6.2a.** *A cold gas.* **6.2b.** *A hot gas. (Notice that the physicist's usage of the word "hot" corresponds to colloquial usage as in "a hot party.")*

A somewhat subtle point here is that the walls are not essential to the argument. Consider a volume of gas inside the dotted line in Figure 6.3, far from the walls of the container. As the gas expands, the molecules near the wall lose energy. While the molecules inside this volume are not aware of the walls, they are aware that the surrounding molecules have slowed down. Due to collisions with the slower molecules, the molecules far from the walls also slow down.

Thus, as Friedmann went up, the air around him expanded and he got colder. Add wind chill, and pneumonia set in. We have here a physicist's postmortem of a cosmologist's death. What does it have to do with the universe?

SOME LIKE IT HOT

Expanding gas cools. By the same physics, the universe also cools as it expands.

It follows that the universe was hotter yesterday than today, and hotter still the day before. As we go back in time, the universe gets hotter and hotter. If we keep going, it would get infinitely hot at some point: We would have reached the big bang.

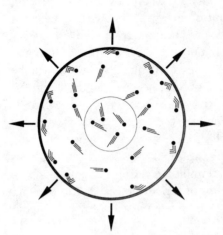

6.3. *Walls are not essential to show that an expanding gas cools. Here we represent a gas expanding in a spherical container. The gas molecules near the container wall lose energy by bouncing off the wall and pushing it outward as indicated. While the gas molecules inside the mathematical volume indicated by the dotted line are not directly aware of the wall, they are aware that the gas molecules around them are losing energy. In an actual gas, the number of molecules is vastly larger, of course.*

CHARACTERISTIC ENERGIES

We live now in a frigid universe: It is about 270° C below freezing out there.

We are talking of the average temperature of the universe, of course. Fortunately for us, there are hot spots here and there called stars, and we have staked out a claim to one of them.

In the past, however, the universe was hot. To get a feel for what life was like in the hot universe, we have to learn about the characteristic energies for various physical processes. By understanding characteristic energies, we will know what to expect at each epoch in the universe's history.

When we witness a blast of dynamite—on film, hopefully—we are shaken by the enormous energy released. But the characteristic energy of the explosion, defined as the energy released divided by the number of particles participating, is actually minuscule and about the same as the characteristic energy involved when a match is struck. All chemical processes—whether burning TNT or the phosphorus in a match—involve more or less the same characteristic energy, give or take a factor of ten or so.

The characteristic energy of chemical processes is the energy involved when electrons skip from one atom to a neighboring atom. If we want to rip an electron off an atom, considerably more energy is needed. As a benchmark figure, physicists use the energy needed to rip the electron off a hydrogen atom, an energy of about 13.6 electron volts. (An *electron volt*, abbreviated as eV, is the amount of energy acquired by an electron when pushed by a voltage of one volt.) The precise amount of energy needed to rip an electron off an atom depends on the type of atom and on how many electrons have been ripped off already, but it will be comparable to 13.6 eV: It may be many tens, or even hundreds, of eV, but it will not be millions of electron volts. Physicists say that the characteristic energy of atomic physics is about 10 eV.

An atom consists of a bunch of electrons orbiting around a nucleus. The nucleus in turn is made of a number of protons and neutrons stuck together. An energy of about 10 million electron volts, abbreviated as 10 MeV, is required to rip a proton or a neutron out of the nucleus. Physicists think of the characteristic energy of nuclear physics as about a few MeV.

Going farther up the energy scale, we come to particle physics, whose characteristic energy is about a thousand MeV, or a billion eV, abbreviated as 1 GeV (giga–electron volt). For instance, if we accelerate a proton until it has a few GeV of energy and smash it into another proton, new types of particles are produced.

This orderly arrangement of characteristic energies represents a kindness on Nature's part. Thus an atomic physicist, when studying processes involving energies of a few eV, need not worry about effects due to nuclear and particle physics, while to a particle physicist smashing a proton with energy of several GeV into an atom, the fact that the electrons in the atom are orbiting around the nucleus hardly matters.

Indeed, a physicist going backward in time through the universe relives a curriculum in physics. He has in succession a course in atomic physics, in nuclear physics, and in particle physics.

Knowing the universe's expansion rate and its present temperature, physicists can easily compute the universe's temperature at any given time. In fact, cosmologists use the universe's temperature as a clock. Instead of saying so many years after the big bang (after all, as a measure of time, a year hardly has any intrinsic cosmological significance), cosmologists often date historical events by the universe's temperature at the time. Recall that the temperature of the universe is just the average energy of the particles contained in it. Thus, when we characterize an epoch in the universe's history by its temperature, we know the average energy of the particles at that time and hence what physical processes are relevant. For instance, when the universe had a temperature of a few MeV, nuclear processes but not atomic processes are relevant.

Depending on how hot the universe is in any given epoch, different physical processes come into play. If we understand the physics that is relevant in each epoch, we understand the universe's history.

This is in essence the insight of the great Russian-American physicist George Gamow. The point seems clear enough, but before Gamow no one had taken the big bang seriously enough to study the physical processes occurring in the early universe. Instead, attention was more focused on finding different solutions of Einstein's equation. Gamow realized that by working out the consequences of these processes, he could demonstrate the reality of the big bang.

THE BOUQUET OF OCTOBER

As a bright young student in Russia in 1924, George Gamow was fascinated by Friedmann's expanding-universe solution. The lad from Odessa was eagerly hoping to work with Friedmann, but the cosmologist's unfortunate death ruined his plans. Instead, another professor inherited Gamow and set him to work analyzing the motion of a pendulum swinging with finite amplitude. Gamow found the problem excruciating and made almost no progress.

To make matters worse, just at that time the Commissariat of Education decreed that all students must pass an examination on "dialectical materialism," the philosophy used by Marx, Engels, and Lenin to prove the correctness of communism. Gamow couldn't make heads or tails of the subject and went into the examination fearing that his physics career might soon be over. "What is the difference between humans and animals?" the examiner asked him. Gamow recalled later that he barely kept himself from replying that humans have souls while animals do not, an answer that would have earned him an instant failure. Instead he replied, "None."

"Wrong!" said the examiner. "According to the book here, humans use tools while animals do not."

But Gamow argued that gorillas were known to beat their enemies with clubs.

Nowadays, thanks to ethologists like George Schaller, who spent years living among gorillas, we know that Gamow maligned these poor, gentle beasts. But he had the right notion: Chimpanzees do use twigs to pick termites out of their burrows. In any case, the examiner was so cowed by the idea of club-swinging gorillas that he passed Gamow and allowed him to continue his studies in physics. Dialectical materialism is wrong on at least one point.

Struggling with the pendulum and dialectical materialism, Gamow was feeling discouraged. Fortunately, his lot took a turn for the better: He was chosen to spend the summer of 1928 in Germany to study the newfangled quantum physics.

The bright young man immediately made his mark by applying quantum physics to the study of atomic nuclei. He returned to the Soviet Union hailed as a hero. The headlines proclaimed, "A Soviet fellow has shown the West that Russian soil can produce her own Platos and sharp-

witted Newtons." *Pravda* even published the following effort by a government poet:

> *The USSR has been labeled the land of the yokel. . . .*
> *Quite right! And we have an example in this Soviet fellow*
>   *named Gamow.*
> *Why, this working-class bumpkin, this dimwit . . .*
> *He went and caught up with the atom and kicked it about*
>   *like a pro.*
>
>   *. . .*
>
> *So the Riddle of Riddles was solved by our poor little*
>   *commonplace nation!*
>
>   *. . .*
>
> *This Gamow is surely a caution. (Take warning, ye lands of*
>   *the West!)*
> *. . . the message is graphic and sober:*
> *In science right now there arises the potent bouquet of*
>   *October.*

ESCAPE FROM DIALECTICAL MATERIALISM

The hero's welcome faded quickly, however. Dialectical materialism soon rose to dizzying heights, and many intellectuals found the atmosphere in Stalinist Russia unbearably stifling.

Gamow tried to escape. In the summer of 1932, he and his wife
attempted the virtually impossible feat of *rowing* across the Black Sea to
Turkey in a kayak they managed to get their hands on. Unfortunately or
fortunately, a freak storm blew them hopelessly off course. After two days
and two nights at sea, soaked and exhausted, they were lucky to escape
with their lives. On another occasion, they actually contemplated skiing
across the Arctic to Norway.

Finally, in 1933, Gamow was able to worm his way out of the
Soviet Union with the help of his physicist friends in the West. They
persisted in inviting him to important international conferences where he
alone had the stature of representing the Soviet Union. The Stalinist
government, torn between its fear of defection and its desire to show the
West the "potent bouquet of October," finally decided to let Gamow go.

Gamow also got his wife out by doggedly refusing to go unless she accompanied him. Later, when it became clear that Gamow was not coming back, an angry Stalin ordered prominent Soviet scientists to be detained permanently.

From Europe, Gamow made his way to the United States. During the war he worked for the military but was not cleared for the atomic bomb project, presumably because of his Russian origins. In 1948, when Gamow was on the faculty of George Washington University, his thoughts turned again to cosmology. The man whom circumstances forced to study the pendulum instead of the universe turned the notion of the big bang into a substantive theory.

BLASTED WITH EXCESS OF LIGHT

He pass'd the flaming bounds of place and time:
The living throne, the sapphire-blaze,
Where angels tremble while they gaze,
He saw; but blasted with excess of light,
Closed his eyes in endless night.

—THOMAS GRAY, *The Progress of Poesy*

Yesterday was hotter than today by about 10^{-11} percent. It wasn't hotter by much, but a billion years ago, the universe was hotter by 7 percent.

Imagine that someone had filmed the universe's evolution. Let us play the movie backward. Millions of years pass in an instant. The heat grows. Soon the universe is glowing. The red glow turns white-hot. All solid and liquid matter is vaporized. The sweet and gentle light that illuminates the world turns murderous. Photons turned muggers threaten the existence of matter. When the average energy per particle gets to be more than a few electron volts, atoms cease to exist. Energetic photons crash into them and rip them apart.

You may wonder why I say a few electron volts, if I just said a while ago that 13.6 eV are needed to rip the electron off a hydrogen atom. Doesn't the temperature have to reach 13.6 eV before atoms are ripped apart?

The point is that by specifying the average height of the population, for instance, we do not imply that everyone would have that height. When the temperature is a few electron volts—that is, when the average energy of the particles is a few electron volts—some of the particles actually have energies in excess of 13.6 eV. Furthermore, the universe contains vastly more photons than protons and electrons. Thus, even if only a tiny fraction of the photons have energies in excess of 13.6 eV, through their sheer numbers they can terrorize the atoms out of existence.

At this point when photons tear atoms apart, the temperature of the universe, by definition equal to the average energy of the particles in it, is a few electron volts. If you prefer more conventional units, one electron volt corresponds to about ten thousand degrees Celsius.

A PLACE OF AGITATION AND VIOLENCE

As we go farther back in time, the universe gets even hotter. When the average energy gets to about one tenth of an MeV, nuclei are ripped apart. (Again, when the average energy of the particles is only one tenth of an MeV, quite a few photons already have energies in excess of a few MeV.)

The universe was too hot for nuclei to exist. A hot caldron of protons, neutrons, and electrons, the universe was full of agitation and violence. Here a rampaging photon smashed into a proton, there a proton careened off a neutron. The protons and the neutrons had too much energy to stick together. Once in a while, by chance, a particularly sluggish proton met an equally slow neutron and they would stick together to form a nucleus called the deuteron. But no sooner had the deuteron formed than a raging photon would come charging along and bust the deuteron up into a proton and a neutron.

Now run the movie forward. We watch the universe cooling down. Particles move slower and slower. It takes about 2 MeV to break up a deuteron. Soon photons no longer have this much energy, and the deuterons have no more fear of being mugged. The number of deuterons increases rapidly. Violence fades as the universe ages.

As the deuterons drift around, some of them are hit by protons and neutrons. When a proton sticks to a deuteron, the resulting nucleus is known as helium 3. When a neutron sticks to a deuteron, the result is a tritium nucleus. The details and the nomenclature do not concern us.

The net result is that as the universe cools, protons and neutrons get stuck to each other. The primeval soup of protons and neutrons is converted into nuclei of various kinds, laying the foundation for the modern world, as a historian might say.

Primeval nucleosynthesis—as the construction of atomic nuclei in the early universe is called—is easy to understand in broad outline. Dab some glue on some marbles and place them in a tray. Begin by shaking the tray vigorously and slow down gradually. At first, a pair of marbles that happen to stick together would almost immediately be ripped apart again as they thrashed about. But as the agitation died down, the marbles would end up sticking to each other in various configurations. (See Figure 6.4.)

YLEM

In 1948, the government had declassified nuclear reaction rates—information about how readily a proton or a neutron would stick to a given nucleus to form a larger nucleus. Gamow realized that with this information he could calculate the relative abundance of the elements in the universe.

6.4

The idea is easy to understand in terms of our tray of marbles. Suppose we start with 120 marbles. In the end, we might find, say, 50 marbles still unattached, 40 marbles bonded together to form 20 pairs, and 30 marbles bonded together to form 10 triplets. The relative abundance of these different configurations obviously depends on the strength of the glue and on the rate at which we slow down the agitation. Use a stickier brand of glue, and there will be fewer marbles left unattached. The strength of the glue corresponds roughly to the nuclear reaction rates, while the rate at which we slow down the agitation corresponds to the expansion of the universe. Gravity, by controlling the expansion rate of the universe, also governs the relative abundance of elements.

In this way Gamow, together with Ralph Alpher, James Follin, Jr., and Robert Hermann, were able to calculate the abundance of helium, tritium, and other elements such as lithium relative to hydrogen in the universe. (Figure 6.5.)

BEFORE THE BROTH CONGEALED

I never saw a moor,
I never saw the sea;

6.5. *Ylem, the first substance from which the elements were supposed to have formed. The genie coming out of the bottle is George Gamow, flanked by his two collaborators, Ralph Alpher and Robert Hermann.*

> Yet I know how the heather looks,
> And what a wave must be.
>
> —EMILY DICKINSON, *Time and Eternity*

The lay reader is often amazed at how physicists could talk of the way the universe was minutes after its beginning with such confidence and vividness, as if they had been there. In fact, it is easier to talk about the early universe than the contemporary universe. The early universe was a homogeneous hot broth. Later, the broth congealed and splattered.

Gamow's logic was impeccable. If we measure in the laboratory what happens when a proton and a neutron, each with an energy of 1 MeV, slam into each other, then we know exactly what happened when a proton and a neutron with the same energy slammed into each other in the early universe. More generally, if we study and understand the physics at about 1 MeV, then we know what happened when the universe had a temperature of 1 MeV.

AFTER THE HOT PARTIES

Gamow thought that all atomic nuclei were formed in primeval nucleosynthesis. But it became clear later that shortly after helium was formed, nucleosynthesis essentially came to a halt. By that time the expansion of the universe had reduced the numbers of protons, deuterons, and helium nuclei per unit volume drastically. The collisions between them were so infrequent that nuclear processes by and large came to a halt.

And thus the universe became a nuclear-free zone, at least for a while. As the universe cooled, the electrons moved ever slower. With the passing of time, the universe became cool enough for the products of nucleosynthesis—the protons, deuterons, and helium nuclei—to grab some passing electrons to form atoms: a proton married to an electron forms a hydrogen atom, a deuteron married to an electron forms a deuterium atom, and a helium nucleus married to four electrons forms a helium atom.

After atoms form, the nuclei can no longer get close to each other. When two atoms get close, the buzzing clouds of electrons keep the two nuclei far apart. No more nights out with the boys after marriage!

THE DULL UNIVERSE

The universe thus settles into a rather dull existence, permeated by enormous clouds of gas. The gas, while hot by human standards, is relatively cool, cool enough for atoms to exist. However, this relatively peaceful nuclear-free state does not last long. Gravity has already been hard at work, pulling together neighboring globs of gas. Soon the first stars condense out of the primeval gas clouds.

As the gas atoms rush together to form a star, they crash into each other with such headlong abandon that they rip electrons off each other. Divorced from the electrons, the nuclei once again find each other and start nuclear-reacting, as nuclei are wont to do. All these nuclear goings-on put out lots of energy. Like a Christmas tree being turned on, the universe is suddenly lit with lights beyond measure.

OUT OF STARFIRES

The details of the nuclear reactions do not concern us—they remind me of high school chemistry class, an experience too painful even to think about. (Add one chemical with an unpronounceable name to another, and what do you get? Yike!) Nevertheless, let us look at one, just one, example of a nuclear reaction that happens to be of great relevance to the human condition. A helium nucleus bouncing around inside a star will bump into another helium nucleus sooner or later. They immediately stick to each other to form a beryllium nucleus. Yet another helium nucleus wanders by and sticks to the beryllium nucleus. The result? A carbon nucleus. Carbon! The cornerstone of the biological world. Out of starfires, we humans become a possibility.

Note the crucial difference between primeval nucleosynthesis and stellar nucleosynthesis. In the primeval setting, nuclei were drifting farther and farther apart in the expanding universe. Later, while nuclei were confined inside stars, they were bound to bump into each other. Thus deuterium, helium, and a little bit of lithium were produced in the primeval universe, while nuclei more elaborate than the helium nucleus—let's call them "higher" nuclei for short—were formed later in stars.

When some of the first-generation stars exploded, they ejected into space higher nuclei, among other things. Out of this ejected debris a second generation of stars soon condensed. These stars started out containing higher nuclei like carbon, out of which more and more complicated nuclei are manufactured. Eventually, these stars in turn exploded and splattered themselves over the cosmos.

SELF-IMPROVEMENT COURSES

You can't make much out of only hydrogen and helium, but with carbon, silicon, iron, and so forth, the possibilities become endless. You can make rocks, for instance. Bits of rocks come together to form rock piles, laughably minute specks of dust in the cosmic scheme of things. On one of these specks, carbon atoms started connecting up with hydrogen, oxygen, and so forth. Somehow—I have no idea how—these bunches of atoms suddenly came alive. Eons and lots of self-improvement courses later, this moving, eating, reproducing ooze turned into what Pogo called human beans.

These guys are pretty clever and discover soon enough the secrets of the fires that forged them. Thus began the third nuclear age. The first nuclear age ended a few minutes after the big bang; the second started with the stars and is still going on; and now we, too, can split nuclei and glue them together. Does anybody care if we blow *our* speck of dust up?

FROM STARDUST TO HOUSEHOLD DUST

Look at the back of your hand. It is covered—as is any other part of your body, for that matter—by a layer of dead cells, strangled by fibrous keratin produced by the cells themselves as they come to the surface. Examined closely, the cells are an intricate latticework of atoms, dominated by carbon, hydrogen, and oxygen atoms. The nuclei of most of these atoms came from deep inside ancient stars that had long ceased to be. We are largely stardust, with a few atoms here and there claiming an even more ancient lineage. Some of the hydrogen atoms might have been around when *star* was a concept unknown to the universe. In your body there are surely even a few deuterium and lithium atoms with their nuclei manufactured in days primeval.

Soon that outer layer of your epidermis will fall off and contribute to household dust—from stardust to household dust.

PALE SHADOW

In their 1948 paper, Gamow, Alpher, and Hermann proposed a crucial test of the big bang. They suggested that a pale shadow of the radiation that filled the early universe should still be visible.

A photon is a packet of electromagnetic waves: Its energy is proportional to the characteristic frequency of the electromagnetic waves it contains. As the universe expands, the energy carried by a photon gradually decreases. The phenomenon may be understood as another manifestation of the redshift. Consider a photon emitted from a distant point. As we learned in Chapter 5, the farther the distant point is from us, the faster it is moving relative to us, and hence the larger the frequency reduction.

This reduction of the photons' energies may also be understood roughly as follows. Momentum measures the ability of a particle to get from here to there. If space itself is expanding, then the momentum of a given particle will decrease—it seems to get harder and harder to get from here to there. Picturesquely, you may think of the photons getting tired.

Gamow and his collaborators showed that the present energy of these primeval photons could actually be determined.

COOKING IN THE EARLY UNIVERSE

A gourmet tastes a hollandaise sauce and mutters disapprovingly, "A touch too much lemon, I say." Physicists proceed along the same line, "tasting" the universe to find out what the Ultimate Cook put in.

By studying the composition of stars and interstellar gas clouds, astronomers have determined the relative abundance of hydrogen, deuterium, helium, lithium, and so forth. The numbers of protons and neutrons relative to the number of photons had to be just right to produce the observed relative abundance of these elements, in exactly the same way that the relative amount of lemon to the amount of egg yolk had to be just right for the hollandaise to taste right. Thus, from observations physicists can deduce the number of photons relative to the number of protons and neutrons in the early universe. The imagery of cooking is apt. In baking, chemical reactions rearrange molecules. In the early universe, nuclear reactions bonded protons and neutrons.

Reasoning along this line and using the known expansion rate of the universe, Gamow and his collaborators estimated what the energies of these primeval photons ought to be now. They had been shifted down enormously in the electromagnetic spectrum, all the way from ultraenergetic gamma rays to microwave electromagnetic waves. A photon containing a packet of microwaves carries about a hundred thousand times less energy than a photon carrying light waves.

THE SONG OF THE UNIVERSE

These microwave photons were all that is left from the once-mighty photons, a faint glow reminding us of the fiery birth of the universe. These telltale photons were detected finally in 1965, totally by chance, at the Bell Telephone Laboratories at Holmdel. There, amid the pastoral splendors of New Jersey, Arno Penzias and Robert Wilson built an ultrasensitive antenna for communication purposes. To their dismay, it produced a steady hum, despite their best efforts, which included, incidentally, periodically removing deposits left by some pigeons who took a liking to the antenna. This brief account does scant justice to Penzias and Wilson, whose dogged persistence was essential in proving that the hum could not be eliminated. They pointed their antenna at the sky in different directions, and always the hum came. In fact, the antenna was picking up the microwave photons left over from the big bang. Penzias and Wilson were listening to the song of the universe.

PORTRAITS IN AGED FRAILTY

Were it not for Penzias and Wilson, we would not even be aware of the existence of the primeval photons as they rained down on us day and night. By now we humans, dazzled by more "modern" sources of photons, the sun and the stars, not to mention electric lights, have long forgotten the once-blinding glare of Creation.

Thus those mighty photons, once able to tear atomic nuclei apart with their bare hands, were reduced to portraits in aged frailty. Drifting through the eons and traveling the lengths of the universe, only to end up in New Jersey. What a fate, eh?

Incidentally, that the detection of primeval photons was not made until seventeen years after the initial prediction, and then only by chance, poses something of a puzzle for historians of physics. The technology was available in 1948. Why then had experimenters not tripped over themselves to look for the glow from the big bang?

Steve Weinberg has given a fascinating analysis of this question. More often than not, history does not develop in a straight line. For one thing, Gamow botched up the details of primordial nucleosynthesis. He arbitrarily supposed that the early universe contained neutrons but not protons. For this and other reasons, the big bang cosmology of Gamow was not taken seriously and gradually faded from the general consciousness. Penzias and Wilson were totally unaware of Gamow's prediction that a faint glow from the big bang ought to be observable.

Quite remarkably, while Penzias and Wilson were working to eliminate the annoying hum in their antenna, not fifty miles from them but unbeknownst to them, a group of physicists at Princeton consisting of Robert Dicke, P. G. Roll, and David Wilkinson were setting up an experiment to detect if the universe had once been hotter. They had also forgotten Gamow's calculation. At Dicke's suggestion, a young theorist named James Peebles worked out primeval nucleosynthesis all over again. He was thus able to predict the expected average energy of the microwave photons. The Princeton group was beaten to the punch. When they heard about the persistent hum picked up at the telephone company lab, they were thunderstruck and immediately realized the magnitude of what Penzias and Wilson had discovered serendipitously.

The song of the universe sounded the death dirge for the steady state theory. The big bang theory predicted the presence of microwave photons left over from the universe's fiery youth. In contrast, there was no compelling reason for a static, unchanging universe to be suffused with microwave photons.

RELIC FOSSILS

Let's now go back to the conversation I had with my friend back in Chapter 4. She had wanted to know what evidence we had for the big bang. I replied that we had fossils. The fossils are the microwave photons and some of the hydrogen, helium, and lithium nuclei. They represent imprints left on the universe by the fire of that primeval epoch.

STRUCTURES OUT OF THE VOID

We look at how different types of matter were born and how a structured world emerged out of the primeval haze.

The Universe Begets Matter

THE BROTH OF MATTER

Nothing can be created out of nothing.

—LUCRETIUS

Out of ylem, a primeval broth of protons, neutrons, and electrons, Gamow cooked up everything.

But where did the ylem come from?

Because you can't make protons and neutrons out of nothing, for a long time physicists regarded this question as unanswerable.

IN CHANGE IS PERMANENCE

Since 'tis Nature's law to change,
Constancy alone is strange.

—J. WILMOT, Earl of Rochester

Plus ça change, plus c'est la memsahib.

—OGDEN NASH

The microscopic world of particles is a world of unceasing change and decay. Whoosh! Here a neutron disintegrates into a proton, a positron, and a neutrino. Kraazam! There a marauding electron crashes into a proton and turns it into a neutron plus an antineutrino. But in change there is permanence. A neutron is changed into a proton, a proton is changed into a neutron, but the total number of protons and neutrons does not change. (The electron, the positron, the neutrino, and the antineutrino are just going along for the ride in these processes converting proton into neutron and vice versa.)

Put six protons and twelve neutrons inside a box. Come back later and look inside. You may find eight protons and ten neutrons, or you may find four protons and fourteen neutrons, but the total number of protons and neutrons is still eighteen and always will remain eighteen. To save themselves from having to say "protons and neutrons" all the time, physicists refer to protons and neutrons (and also some other related particles) as baryons, a word with the same root as "baritone" and meaning "heavy particle."

Our imaginary box will always contain eighteen baryons, not one more, not one less. Physicists refer to this fact as the conservation of baryon number: Physical processes can only change one baryon into another, but they cannot cause a baryon to disappear into or to pop out of thin air. This was first enunciated as a sacred truth in the 1930s by the eminent Hungarian-American physicist Eugene Wigner. (Of course, to mutter "Baryon number conservation" is not to explain baryon number conservation; it merely summarizes observational facts. Nobody has ever seen a baryon disappear.)

PROTONS ARE FOREVER

Baryon number conservation has one important consequence. Picture a proton sitting there. Baryon number conservation says that if the proton could disintegrate at all, it would have to disintegrate into the neutron or some other baryon. But it so happens that the proton is the lightest of all baryons. The neutron, for example, is just a teensy bit (about 1 percent) more massive than the proton. The proton cannot disintegrate into the neutron because the extra mass cannot pop out of nowhere. There is nothing for the proton to disintegrate into. Ergo, immortality! Baryon number conservation, together with energy and mass conservation, guarantee the immortality of the proton.

In contrast, a neutron disintegrates in about ten minutes into a proton, a positron, and a neutrino. Incidentally, that the neutron happens to be slightly more massive than the proton is one of those mysterious but happy accidents that make our world possible. Imagine a universe in which the proton is more massive than the neutron. Then baryon number conservation would guarantee the immortality of the neutron but not that of the proton. The proton would decay into the neutron. Consider the simplest of all atoms, the hydrogen atom. In a hydrogen atom, an electron goes round and round a proton, the electron held in bondage by its electrical attraction to the proton. Suddenly whoosh, the proton turns into a neutron. Hey, what happened? yells the electron. The neutron, being electrically neutral, does not attract the electron. Freed from its bondage, the electron zings into space. Thus, in this imagined universe, the hydrogen atom is unstable, and the world looks drastically different. Why the neutron happens to be more massive than the proton remains unexplained to this very day.

Well, even though we don't understand why, be glad that we live in a universe with an immortal proton rather than an immortal neutron.

FEEL IT IN YOUR BONES

Baryon number conservation accounts for the stability of the proton. We can turn it around and say that the obvious fact that the material world is not disintegrating around us provides good evidence for baryon number conservation. When the distinguished experimenter Maurice Goldhaber was once asked how Wigner knew about baryon number conservation, he joked in reply, "Oh, he could probably feel it in his bones." Indeed, if protons were in fact disintegrating, then we would be disintegrating. Since a photon is emitted whenever a proton disintegrates, we would all be glowing in the dark. (Well, all right, if you want to pick nits, everything around us would also be glowing: We would glow, but not in the dark.)

Of course, physicists cannot guarantee that a proton will last forever. Rather, as practitioners of an empirical science, physicists can assert only that the facts indicate the lifetime of the proton to be at least such and such. Indeed, the very existence of matter implies that the proton's lifetime is longer than the age of the universe, 10^{10} years. Hopefully, protons live a lot longer than that.

Because of the probabilistic nature of quantum physics, physicists speak of lifetime in the same sense that insurance executives speak of lifetime. Suppose that protons live 10^{10} years. Watching a given proton, you would have to wait 10^{10} years on the average before you see the thing go poof. But if you watch a collection of 10^{10} protons over one year, chances are one of them will disintegrate within the year.

Using Goldhaber's quip, we can easily show that protons live a lot longer than the present age of the universe. The point is that a gram of matter contains an enormous number of protons—about 10^{24}, in fact. Were the proton lifetime only (!) as long as 10^{10} years, then statistically, 10^{14} protons in that gram of matter would decay within one year. Knowing how much energy is emitted every time a proton goes poof, we can then figure out how much energy would be emitted per second by a single gram of matter. Just from the fact that we are not all glowing, physicists can state that the proton lives for at least a million times longer than the present age of the universe. This is such a long time that it seemed pointless even to speculate on whether the proton lives forever or not.

NO BIRTH WITHOUT DEATH

Thus, physicists came to believe in absolute baryon number conservation as a law engraved in stone. They abandoned hope of ever understanding where Gamow's broth of protons and neutrons came from. It was just there.

The reason for despair is clear. Just as baryon number cannot vanish into thin air, so baryon number cannot appear out of thin air. If the proton cannot disintegrate, it also cannot be born. Without death, there can be no birth. See, I can even make it sound vaguely like Eastern philosophy.

Suppose you could go out and count the total number of protons and neutrons in the universe. (Astronomers can actually do this. First, they count up the number of galaxies. Knowing how many stars there are on the average in a galaxy and how many protons and neutrons in a star, they know the total number of protons and neutrons in the universe.) After much labor, you determine that there are one gazillion protons and neutrons in the universe.

Heareth, the law of baryon number conservation speaketh! You mere mortals, know that since the baryon number of the universe is now one gazillion, it has always been and always will be one gazillion! Not one more, not one less.

When I was starting out in physics, the total number of protons and neutrons in the universe was said to be beyond the realm of physics, since no physical process can change that number. Whoever started the universe going had thrown in one gazillion protons and neutrons, and that's that. Since there was no hope of ever understanding why this was so, people didn't think about it.

ANTIMATTER

In 1931 the brilliant English physicist Paul Adrien Maurice Dirac astonished the physics world by predicting that, corresponding to each particle, an antiparticle existed. Dirac was no crackpot. Within a year, the antielectron was discovered. The antiproton, however, was not discovered until 1955, when experimenters observed the production of a proton and an antiproton in the collision of two energetic protons—in other words, in this reaction:

proton + proton → proton + proton + proton + antiproton

What about the law of baryon number conservation? It works beautifully, thank you. The antiproton counts as minus a baryon. Thus, in the reaction above, the baryon number started out equal to $1+1=2$ and ended up equal to $1+1+1-1=2$.

When I said that you cannot make a proton out of nothing, you might have thought I meant that the proton is impossibly difficult to make, like an intricate toy. But that is not so. You can make a proton easily enough simply by smashing two protons together. The catch is that you can't just make a proton, you have to make a proton and an antiproton. Making a proton, impossible; making a proton and an antiproton, easy. The subnuclear world is weird.

Instead of saying that we cannot make a proton or a neutron out of nothing, we should say more correctly that we cannot change the baryon number of the universe.

All this talk about antimatter slightly complicates our preceding discussion but does not alter our essential conclusion. If the baryon number is one gazillion, then the universe could have started with one gazillion and seventeen protons and neutrons and seventeen antiprotons and antineutrons. Or perhaps seven gazillion protons and neutrons and six gazillion antiprotons and antineutrons. The net baryon number is, however, always one gazillion.

Incidentally, the existence of antimatter has long been established. At big accelerators, experimental physicists see pairs of particles and antiparticles come into being all the time, before breakfast, after lunch, whenever.

Antimatter obeys the same physical laws as matter, as I will discuss in more detail later. An antiproton and an antielectron would make an antihydrogen atom. In this context, the preface "anti" does not carry any pejorative sense whatsoever. We call antimatter anti merely because we knew about matter first. Antimatter is just as valid as matter, and physicists can easily imagine a universe made of antimatter rather than matter, or a universe populated equally with matter and antimatter. (In fact, we will come to this second possibility shortly.) In this chapter we have been pursuing the question of where the matter in the universe came from. This question is essentially the same as how the universe came to contain matter rather than antimatter.

QUARKS

Look, you say, why all this talk about the impossibility of making the protons and neutrons out of nothing? Physicists now know that the proton and the neutron are each made of three quarks.

But that just begs the question: Where did the quarks come from? The law saying that you can't make protons and neutrons out of nothing just gets translated into a law saying that you can't make quarks out of nothing. Baryon number conservation is equivalent to quark number conservation. If the universe's baryon number is one gazillion, then its quark number is three gazillion, but we still have no hope of understanding why the quark number is three gazillion and not three gazillion and one.

Nothing profound here. All I am saying is that whether you prefer to think of matter as consisting of protons and neutrons or of quarks, the same question of where matter came from remains.

AN UNCLUTTERED UNIVERSE

How cluttered is the universe? You counted that the universe contains one gazillion protons and neutrons. Well, is that a lot, or a little? To what are we to compare one gazillion? The national deficit in dollars?

Before the discovery of cosmic microwave radiation in 1965, there was no way of answering this question sensibly. But with that discovery, physicists knew how many photons the universe contains. It is natural to compare the number of baryons to the number of photons, and it makes sense to ask: How cluttered with matter is the universe?

Well, it turns out that there are ten billion photons for every baryon. Matter is a one-part-in-ten-billion contamination in an otherwise pristine universe. I like to think of matter as the dirt in the universe. To fundamental physicists, a universe devoid of matter appears pure and elegant, filled with nothing but the sweetness of light.

Looking up at the dark night sky, we feel with Pascal that there are far too few stars compared to the vastness of space. The number ten billion measures this almost inconceivable emptiness.

Viewed in this light, it is rather curious that Whoever threw in the dirt decided to throw in just a minuscule amount. Indeed, why would He want to throw in any dirt at all?

WORLDS AND ANTIWORLDS

Faced with this conundrum, some physicists, notably the French physicist Roland Omnès and his colleagues, concocted a nifty solution: They did not throw in any dirt at all. Having no hope of understanding the number 0.0000000001, which measures the amount of dirt relative to light, they boldly asserted that the number is actually 0. There is no dirt.

The idea is that the universe could start without any baryons at all. Pairs of baryons and antibaryons could then be produced by the collisions of the particles that were present. No matter how complicated these production processes might be, absolute baryon conservation guarantees that there always would be an equal number of baryons and antibaryons. The matter and antimatter got segregated, somehow, into different domains as the universe evolved.

111

According to this view, we are wrong to think that the entire universe is constructed out of matter just because our immediate neighborhood is filled with matter. Perhaps the galaxy next to ours is made of antimatter.

The scenario that the universe is divided into matter and antimatter domains holds enormous appeal for science fiction writers. Think of the dramatic possibilities! But alas, it does not hold up under scrutiny.

Observationally, one might expect to see some antinuclei in cosmic rays, interlopers from another domain. But they have not been seen. One might also expect that at the boundary between two domains, matter and antimatter would be annihilating furiously, emitting extremely energetic photons. Again, astronomers have not detected these telltale photons. Furthermore, the proponents of this scenario never succeeded in finding a convincing mechanism that would segregate matter and antimatter. These difficulties proved to be insurmountable, and so the matter-antimatter scenario was rejected.

In the summer of 1974, my wife and I went to Paris and happened to stay in the apartment of a collaborator of Roland Omnès. Research physicists often travel to other institutions in the summer and usually end up staying in apartments and houses belonging to physicists who are themselves away on trips. It was a cozy apartment, with shutters opening onto a tiny balcony from which we could observe Paris going by. The other thing I remember is that lying around were some papers by Omnès and his collaborators on the matter-antimatter universe.

At that time I was working on problems far removed from those of cosmology. As I said, physicists had convinced themselves that there was no hope in understanding where the baryons came from, so they didn't think about it. Omnès's ideas, meanwhile, had long been discredited. Fundamental physicists were excitedly exploring the newly established theories of the microscopic world. Cosmology in general and the origin of the baryons were not in their collective consciousness. I never heard cosmology mentioned during my graduate school years at all.

Nevertheless, those papers caught my eye. I glanced at them and filed them away in my head. Some two years later, back at Princeton University, where I was on the junior faculty, I talked to my colleague Frank Wilczek about the problem of where baryons came from, and we spent some time trying to invent a way of segregating matter and antimatter in the early universe. Given the then newly established theories of the microscopic world, we thought we had a good chance of success where Omnès had failed. We didn't get anywhere, so we dropped the project. All along, we were still worshiping at the altar of absolute baryon conservation.

GRAND AND UNIFIED

Meanwhile, a grand unified theory of the fundamental particles was proposed in 1974 by Jogesh Pati and Abdus Salam and independently by Howard Georgi and Shelly Glashow. Let us digress briefly to understand the nature of the grand unified theory.

Modern physics identifies in Nature four fundamental forces: the electromagnetic, the gravitational, the strong, and the weak. The electromagnetic force holds atoms together, governs the propagation of light and radio waves, causes chemical reactions, and prevents us from sinking through the floor. The gravitational force keeps us from flying off into space, holds planetary systems and galaxies together, and controls the expansion of the universe. The strong force holds the quarks together in subnuclear particles such as protons and neutrons, and the weak force causes certain radioactive nuclei to decay. Although the strong and the weak forces are essential to Nature's design, they do not play any role in any phenomena on the scale of everyday life.

For a long time, physicists thought of these four fundamental forces as unrelated. Then in the early 1970s the suspicion grew steadily that in spite of their apparent dissimilarities, three of the four fundamental forces—the strong, the electromagnetic, and the weak—are related at a deeper level. In the grand unified theory, these three forces are merged into one grand unified force, with a characteristic energy scale of about 10^{15} GeV. The grand unified theory predicts that in processes in which particles with energies of about 10^{15} GeV collide with each other, the strong, the electromagnetic, and the weak forces would become the same force. Recall from Chapter 6 that the characteristic energy scale of particle physics is a few GeV and that of nuclear physics a mere few MeV— that is, a few thousandths of a GeV. The characteristic energy of grand unified physics is mind-boggling in comparison.

Considering that the world's largest accelerators have accelerated particles only to energies of several hundred GeV, physicists will not be able to verify the grand unified theory in the laboratory for a long, long time to come. They can only hope to check the theory indirectly.

The story of how physicists came to unify three of the four fundamental forces is without doubt one of the most exciting in the history of physics. I told this story in considerable detail in my book *Fearful Symmetry*. Here I have to limit myself reluctantly to one of the most striking indirect predictions of the theory.

THE MAGICIANS OF CHANGE

Before grand unification, I could take a piece of paper, draw a line down the middle, and write down the names of all the quarks, the tiny particles that make up protons and neutrons, on the left side of the line, and the names of all the leptons, a generic term for particles that resemble the electron, on the right side of the line. The fundamental forces act on these particles, the quarks and the leptons. These forces can also transform one particle into another. For instance, the strong force transforms quarks of one type into another but leaves the leptons alone. Out of these transformations emerge the splendors of the material world.

But in all these transformations, a particle on one side of the line is never changed into a particle on the other side. The particles on the left side of the paper, the quarks, and the particles on the right side, the leptons, live in their separate worlds. In *Fearful Symmetry* I spoke of a magician whose art is limited to transforming one animal into another animal, one fruit into another. A rabbit and an apple are on the stage. The magician waves his cape, and, *whoosh,* the rabbit and the apple are transformed into a fox and some sour grapes. The audience bursts into applause. *Whoosh,* the fox and the grapes are gone, replaced by a mouse and a watermelon. But no matter how fantastic the transformations, there always will be one animal and one fruit onstage. So, too, the fundamental forces can only transform one quark into another quark, one lepton into another lepton.

You may recognize that what I have just told you implies baryon conservation: The three quarks that made up a baryon can be transformed only into three other quarks. There always will be three quarks, just as there always will be one animal onstage in the analogy.

Grand unification erases the line I have drawn on the piece of paper. The worlds of quarks and leptons can no longer be separated; the two worlds are unified. With grand unification come new transformations that change quarks into leptons and vice versa.

Onto the stage struts a new magician. Applause and *whoosh*, the rabbit is transformed into an orange. No more animal onstage. So, too, in grand unified theory, baryon number is no longer conserved, and the proton becomes mortal. The three quarks inside a proton can be transformed into leptons such as electrons or antielectrons. *Whoosh*, no more baryon onstage.

DIAMONDS ARE NOT FOREVER

The theory predicts the fall of baryon number conservation and that the proton will disintegrate into an antielectron (also known as a positron) and an ephemeral particle known as a pion.

The initial reaction of physicists was that a theory with such an outrageous prediction could be ruled out easily, since the proton lifetime was already known to be a lot longer than the present age of the universe merely from the fact that the world was not glowing. That's such a long time, they thought, surely the theory would predict a much shorter lifetime and hence could be trashed immediately.

Well, when the calculation was done, the proton's lifetime came out to be about 10^{30} years. The theory cleared its first hurdle. This is quite remarkable, since a priori, before one plugs the numbers in, one wouldn't know that the proton's lifetime might not come out to be 1 second, 1 year, or even 10^{10} years. The theory could easily have failed, but it didn't.

The mind reels before a time scale like this, a time scale that makes an eon look like a wink of the eye. I cannot really grasp how long ago the dinosaurs roamed the earth, so how can I understand the lifetime of the proton? I can't. I can only throw some numbers at you. The universe is 10^{10} years old. There are about 3×10^7 seconds in a year. Thus the universe is about 3×10^{17} seconds old. Think of all the seconds that have ticked by since the universe began. Now imagine expanding each second into the age of the universe. The total time elapsed would still be "only" $3 \times 10^{17} \times 10^{10} = 3 \times 10^{27}$ years, three hundred times shorter than the predicted lifetime of the proton. The proton lasts a long time.

The mortality of the proton can be readily tested, at least in principle. Just follow Wigner's quip: Gather up a large pile of matter and watch it carefully for telltale bursts of light. Since water is the least expensive material transparent to light, experimenters have taken to watching over large tanks of water for long periods of time. (Figure 7.1).

Unfortunately, experimenters have yet to see any protons dying. In fact, they have now established that the proton's lifetime has to be at least 10^{32} years or so. Thus the original version of the grand unified theory has been ruled out. Nevertheless, because of the beauty and elegance of the theory and because the theory did have a successful prediction about the weak interaction, many physicists—but by no means all—continue to believe in the grand unified theory. There are modified versions of the grand unified theory in which the proton's lifetime comes out to be longer than 10^{32} years.

What I am going to tell you next is predicated on the assumption that the grand unified theory is correct and that the proton indeed does not live forever. When I have to mention a number for the proton's lifetime, I will use 10^{32} years just to be specific.

7.1. *Looking for a disintegrating proton.*

TO COOK AN EGG

Some popular expositions of physics give the impression that one day the grand unified theory was proposed, and the next day physicists were all thinking about it. This is far from the truth. Grand unified theory was proposed in 1974, but physicists did not embrace it immediately. For one thing, at that time even the theory of the weak force, on which the grand unified theory was based, was very much in doubt. The theory contradicted certain experiments that since have been proven wrong. Physicists were not ready to accept the theory of the weak force, let alone a theory that purported to unify the strong, the electromagnetic, and the weak forces. To many, the grand unified theory seemed like so much wild speculation. Psychosocial factors, often ignored by popular would-be historians, also played a role. When the grand unified theory first came out, I heard that a distinguished senior physicist dismissed the theory curtly: What nonsense, a theory that makes the proton decay! Notions entrenched over a lifetime in physics—in this case baryon number conservation—are extremely difficult to shake.

In any case, in late 1977 and early 1978, Frank Wilczek and I, together with two Princeton colleagues, Doug Toussaint and Sam Trieman, thought on and off about the origin of matter in the universe. We recognized baryon nonconservation to be the key, but we were still reluctant to think in terms of the grand unified theory. Other physicists were also thinking about the problem, but again not in terms of the grand unified theory.

The Japanese physicist M. Yoshimura was the first to suggest that the grand unified theory could account for the presence of matter in the universe. A number of physicists then fleshed out this suggestion. At first sight it might appear that the idea obviously would not work. If it takes 10^{32} years for a proton to die, it may also take 10^{32} years for a proton to be born. This is so because in particle physics the process by which a particle is born is just the reverse of the process by which the particle dies, and the rates of the two processes are related.

Here Gamow's insight as described in the preceding chapter comes in. As we go back into the early universe, the temperature rises higher and higher. The characteristic energy of the grand unified theory, 10^{15} GeV, is positively mind-boggling in its enormity; yet, if we keep going back, we eventually will reach a point when the universe was as hot

117

as 10^{15} GeV. At this temperature, particles were crashing into each other like freight trains. The strong, the electromagnetic, and the weak forces were merged into one, and the processes generating baryons occurred rapidly.

A process that takes 10^{32} years at low temperatures takes a zillionth of a second at high temperatures. Sound familiar? You bet! It is one of the basic principles of cooking: Use heat. Consider boiling an egg. The complicated chemical processes that turn a raw egg into a hard-boiled egg would take a long time to occur at room temperatures—indeed, other chemical processes responsible for turning the egg into a stink bomb would occur much sooner—but the same processes occur within a few minutes at the boiling temperature of water. The other side of the coin is, of course, the use of refrigeration to retard chemical processes.

NATURE PREFERS MATTER

Wait a minute, you say, something smells fishy here, and it is not the rotten egg we just talked about. It is all very well that baryon number nonconserving processes went at a clip in the extremely early universe. But how do these processes know enough to generate baryons and not antibaryons?

Good point! In Dirac's theory, the equations governing antimatter turn out to be exactly the same as the equations governing matter. Let us express this striking equality in physical terms. Take any process involving matter, and watch how things go. Now consider the same process but with matter replaced by antimatter. Then, because the equations governing antimatter are exactly the same as the equations governing matter, this process, with matter replaced by antimatter, will go in exactly the same way as before.

Physicists say that Nature does not prefer matter over antimatter, nor does She prefer antimatter over matter. This is known in the trade as CP conservation. (CP stands for charge conjugation and parity.)

What does CP conservation imply for the genesis of matter? We were happily cooking up baryons in the early universe, talking gleefully of processes that would generate baryons. But that was because we didn't know about CP conservation.

CP conservation tells us that the broth won't come out right. Any processes that generate only baryons would favor matter over antimatter, and that is not allowed by CP conservation. CP conservation requires that these processes must also generate antibaryons some of the time. Indeed, it requires that on the average, these processes must generate an equal number of baryons and antibaryons. As long as CP conservation holds, we can again despair of ever understanding how the universe came to be filled with matter, even if baryon number conservation is violated.

Why do physicists believe in CP conservation? Is it because Dirac said so? Of course not. Experimenters had checked CP conservation in process after process and always had confirmed it—until 1964.

In that year, a group of experimenters led by Jim Cronin and Val Fitch discovered that surprise, surprise! Nature favors matter over antimatter by a tiny margin of about one in a thousand in the disintegration of a subnuclear particle known as the K meson. CP conservation was dethroned as an absolute law of Nature. It is almost as if Nature wants to confound those nosy physicists. Almost all of the time, I will show no preference for matter or antimatter, but once in a long while I will favor matter over antimatter in the disintegration of a subnuclear particle.

For a long time after 1964, physicists had not the foggiest idea of how CP violation fit in the scheme of things, considering that it was observed only in the decay of some obscure particle. Then finally, in 1978, this obscure phenomenon emerged as a key player in the early universe. No need to despair! Rejoice—CP conservation has fallen, and we can understand the genesis of matter!

COSMIC ARROW OF TIME

We are finally near the denouement of our story. The final ingredient necessary to cook up matter is an expanding universe. A static universe just won't hack it. It is easy to see why. Consider a process that generates baryons. But the same process reversed removes baryons. Baryons are removed as fast as they are generated, and we are once again left with no baryons.

To be more precise, consider a process in which a particle X decays into a particle Y and a particle Z. In the reversed process, Y and Z would come together to form X. The process and the reversed process would reach an equilibrium in which the effect of the process is exactly balanced by the effect of the reversed process.

How is the situation different in an expanding universe? Ah, there is a big difference. While particles Y and Z are trying to come back together, the universe has been expanding, and so particles Y and Z find themselves farther apart. It gets harder for them to meet to form back into X. In an expanding universe, the forward process dominates the reversed process.

We are really saying the obvious. A static universe is essentially unchanging. For baryons to appear, we need an arrow of time, and cosmic expansion provides that arrow. Gravity controls how fast the cosmic oven is cooled. Too fast or too slow, and we may have ended up with quite different amounts of matter.

Out of barley and mutton and the chemical law of molecular combination we make Scotch broth. Out of primeval energy and cosmic expansion and the violation of baryon number and CP conservation we make the broth of matter, Gamow's ylem. But unlike Scotch broth, which consists of a blending of what there was, the broth of matter consists of what wasn't. Out of this exquisite broth come the splendors of the world.

ST. AUGUSTINE'S ERA

And we suddenly know what heaven we're in
When they begin the beguine.

—COLE PORTER

Before the beginning of years
There came to the making of man
Time with a gift of tears,
Grief with a glass that ran.

—A. C. SWINBURNE

Remember how as a child listening to a fairy tale you always wanted to hear about what happened next? When listening to the tale of the early universe, we naturally want to ask what happened before. Human curiosity is not satisfied until the ultimate end or the ultimate beginning is detailed. What caused the big bang? Who initiated the big bang? What happened before the big bang?

Gamow said, Hey, if we understand nuclear physics, then we

understand the behavior of the universe at a temperature equal to the characteristic energy of nuclear physics. The flip side of that insight is that if we don't understand nuclear physics, then we can't possibly understand the behavior of the universe at the corresponding temperature. Any attempt to do so would really be no better than wild speculation. Gamow had no way of understanding the universe at a temperature one hundred times greater, say. Ask him where his ylem came from, and he wouldn't know.

Having had Gamow as one of my heroes in high school, I was excited to follow in his footsteps. What Gamow's generation did for the genesis of nuclei, mine did for the genesis of matter. A scant thirty years after Gamow, physicists had pushed up the energy scale of their understanding enormously. Now that they have the hubris to claim that they know something about physics at a characteristic energy scale of 10^{15} GeV, some 10^{18} times the energy scale of nuclear physics, they can discuss the genesis of matter. Keep in mind, however, that the grand unified theory is not experimentally established, as nuclear physics was in Gamow's time. Physicists can now say with confidence that the relative abundance of helium to hydrogen has to be such and such, but they are quite unable to calculate the relative abundance of baryons to photons. They don't understand the violation of baryon number and of CP well enough.

In this context, an interesting question is whether we can predict that the universe is made of matter rather than antimatter. Here is a test to see if physicists really understand the physics involved in the genesis of matter. Put some physicists in a lab. Let them observe K mesons decay. Give them all the paper and pencils they want. Can they, without ever looking out the window of the lab, predict what the world outside is made of?

At the moment, they can't. The problem is to relate the observed CP violation in K meson decay to the supposed CP violation in the baryon-generating processes.

Again, the obvious. We know the physics at the nuclear energy scale, so we understand the universe at the nuclear temperature—that is, when the universe has a temperature equal to the characteristic energy of nuclear processes. We know the physics at the particle energy scale less, so we understand the universe at the particle temperature less well. We think we know a little something about physics at the grand unified scale, and thus all we can offer is a rough outline of the universe at that time.

Now, what were you asking? Oh, yes, you wanted to know what happened before the genesis of matter. The answer is that we don't know. Without a firm understanding of the physics at arbitrarily high temperatures, physicists can only speculate. Yet there are those who talk of the ultimate, the moment of the big bang, and what led up to the big bang. It is hype—nothing more than wild guesses.

Once, Gamow was asked what happened before the big bang. He answered that physicists know that period in the universe's history as St. Augustine's era. Why? St. Augustine of Hippo first asked, "What was God doing before He made Heaven and Earth?"

Incidentally, St. Augustine suggested an answer: "He was preparing Hell for people who pry into mysteries," but immediately dismissed it as frivolous. He concluded that "before God made Heaven and Earth, He made nothing." He then went on to say, addressing God, "If there was not time before Heaven and Earth were created, how can anyone ask what You were doing 'then'?"

The Rich Get Richer

The universe is a spiraling Big Band in a polka-dotted speak-
easy, effectively generating new light every one-night stand.
—ISHMAEL REED

TRADING IN THEIR FREEDOM

In the history of the universe, the formation of galaxies was a
particularly momentous event. Before that, the universe was an undiffer-
entiated, uniform cloud of matter. But about forty thousand years after
the big bang, bits of matter started to come together, forming enormous
structures that eventually condensed into galaxies. Galaxy formation
marked the first step in the emergence of structures in our universe:
Within the galaxy, protostars soon formed. The rest, as they say, is history.
The debris around some stars eventually cooled into lumps. On some of
these lumps, biological evolution began. What happened afterward, some
cosmologists might say, is a yawning anticlimax. Certainly, from the
cosmological point of view, nothing much of significance had transpired
since the lumping of matter made the universe take on its present look—a
vast and inconceivable emptiness dotted here and there by a few specks
of matter.

We will look at how structures evolved out of the primeval haze.
Somehow, atoms and the constituents of atoms—electrons, protons, and
neutrons—had come together, trading in their freedom to roam the
cosmos unfettered for a cooperative existence.

THE RICH GET RICHER

Long ago, Newton had already identified the basic physics responsible for the emergence of structure: the inherent instability of gravity.

In 1692, a certain Reverend Dr. Richard Bentley, chaplain to the Right Reverend Father in God, Edward, Lord Bishop of Worcester, gave a series of sermons later published under the title *A Confutation of Atheism from the Origin and Frame of the World*. One of Bentley's theses was that the universal presence of gravity proved the existence of God, a view with which Newton was much in sympathy.

Bentley wrote to Newton with some questions about physics. A lively correspondence followed. In one of his letters to Bentley, Newton suggested how structures could emerge in the universe. I transcribe Newton's argument into modern language as follows:

Imagine space filled uniformly with matter. A mathematician would take that to mean the distribution of matter is perfectly and absolutely uniform. But in an interesting and apt illustration of how physics differs from mathematics, a physicist would naturally consider what would happen were the uniformity not perfect. Newton made the point that any irregularity, no matter how minute, will grow larger.

Consider a region with more matter per unit volume than the surrounding regions. Being denser, the matter in this region would pull matter in from the surrounding regions by the force of gravity. As a result, the matter distribution in this region becomes even denser. The process accelerates—it is the cosmic equivalent of the often-observed phenomenon that the rich get richer and the poor get poorer. "And thus might the sun and fixed stars be formed," concluded Newton. (Galaxies were unknown in Newton's time.)

Newton's scenario makes such obvious sense that it remains the basic explanation of how structures emerged in the early universe. Small fluctuations in the density of matter grew and became amplified.

Incidentally, historians know about Newton's letters because in 1756 Bentley's heirs published them under the title *Four Letters from Sir Isaac Newton to Doctor Bentley Containing Some Arguments in Proof of a Deity*.

A PRIMEVAL CONTEST

In its intrinsic instability, gravity is dramatically different from the other forces. The electromagnetic force, the other long-ranged force in Nature, is intrinsically stable because it acts oppositely on positive and negative charges. To see this, consider a gas consisting of equal numbers of protons and electrons. While opposite charges attract, like charges repel. Thus the electrons want to space themselves apart from each other. Similarly for the protons. An excessive concentration of electrons in one region would be immediately smoothed out by the mutual repulsion between the electrons. Unlike gravity, the electromagnetic force tends to smooth matter out.

Thus far in this chapter, I have talked only about gravity. As you will see shortly, the electromagnetic force also plays a role. In its tendency to smooth matter out, it counteracts gravity's tendency to clump matter together. Galaxies formed and structures emerged from this primeval contest between these two forces.

SLOWED DOWN BY EXPANSION

In 1902, the English physicist Sir James Jeans made the next major contribution to the theory of galaxy formation. He tried to calculate the size of the actual lumps that would form. However, he did not know about the universe's expansion. Hubble was only thirteen at the time.

Clearly cosmic expansion, by thinning out the distribution of matter, works to slow down the formation of lumps. In our analogy, cosmic expansion acts like taxation: As the rich get richer, part of their wealth is continuously removed. But the tax rate is more or less flat: Regions sparse with matter are stretched out at essentially the same rate as the regions dense with matter. A calculation including the effect of cosmic expansion, first done by the Soviet physicist E. Lifschitz in 1946, shows that lumps will still form but at a far slower rate than would have been the case were the universe static. In Jeans's calculation, lumps form exponentially fast, much as how wealth accumulates under compound interest. With taxation, the rich continue to get richer; they are merely slowed down.

DAWNING OF THE ATOMIC AGE

To follow the emergence of structures in the expanding universe more closely, we have to examine matter in more detail. We now know that matter consists of atoms—something that old Isaac didn't know. The hydrogen atom, the simplest atom, is made of a positively charged proton with a negatively charged electron orbiting around it. Recall that in the hot early universe, atoms did not exist: Particles were zinging around so fast that atoms would have been torn apart immediately.

Historians love to talk about crucial events that changed the subsequent course of history. In cosmic history, the formation of atoms is one such crucial event. At some point, the universe had cooled sufficiently for a proton and an electron to come together to make a hydrogen atom.

How did the emergence of atoms alter the character of the universe? To see how, imagine that you were a photon in the early universe before atoms formed. The universe was full of charged particles, protons and electrons, with which photons interacted vigorously. As you zinged around at the speed of light, you kept elbowing protons and electrons aside. Indeed, by crashing into protons and electrons at every opportunity, you and your buddies were making sure that they did not get together to form atoms. (Figure 8.1)

8.1. *In the primeval universe, photons rushed about vigorously trying to prevent the electrons from attaching themselves to the protons.*

Meanwhile, the universe kept expanding—there wasn't much you could do about that. You and your buddies were feeling less energetic by the minute. The place was cooling down. All of a sudden, the electrons and protons were combined into electrically neutral atoms. You guys were too tired to bust them apart.

Your life changed. As you zinged along, you would encounter an atom now and then. Most probably you would pass right by the atom, barely bothering to say hello.

Why this lack of desire to interact with atoms?

In everyday life, we think of photons interacting vigorously with atoms. All around us, photons are bouncing off atoms with zest. In a flash, zillions of photons are absorbed and emitted by atoms. The interaction between photons and atoms could even be violently intense, as anyone hurting from a sunburn can attest. Nevertheless, the interaction of a photon with an atom is feeble compared to its interaction with charged particles. The reason: The photon interacts oppositely with the positively charged protons and with the negatively charged electrons in an atom. The net effect is almost zero. (The effect does not add up to exactly zero because the electrons are spread out while the protons are concentrated in the nucleus.) What we think of in everyday life as a vigorous interaction of photons with atoms is actually, on the scale of fundamental physics, a rather feeble residual effect.

That's why, you the primeval photon, once so ebullient as you careened into things every other instant, now cruise right on by these atoms. Sure, once in a long while you may interact with one of them, but it was not like how it used to be in the good old days. The universe with these newfangled doodads called atoms looked almost transparent to you. From this point on, you just kept on going. Given the vastness of space, the chance is slim that you would bump into a star or a planet, but if you did, you would encounter so many atoms all at once that you would end up interacting with one of them. Who knows, you might even end up in New Jersey in an antenna encrusted with pigeon deposits and set the physics community on its ears.

COSMIC ROBIN HOODS

Let us now look at the primeval contest I already alluded to between the electromagnetic force with its tendency to smooth things out and gravity with its tendency to clump things together.

127

Remember, there are ten billion of you and your fellow primeval photons for every proton and electron. So you guys are the dominant players. Picture the universe before atom formation. As soon as some regions happen to have a higher concentration of protons and electrons, you fellows would push the protons and electrons around and smooth out the distribution of matter. You and your buddies act as sort of cosmic Robin Hoods, making sure the rich don't start to get richer until the ordained starting time. (Figure 8.2.)

Hey, when can we rich types start to accumulate more wealth? When is this ordained starting time, anyway?

The formation of atoms, that's when. The mad scramble for wealth cannot start until atoms form. Suddenly you photons don't interact much with matter. You no longer care about what matter does: Particles can now go out and lump together all they want. Galaxy formation can now start.

In the physics literature, the formation of atoms is said to signal the decoupling of radiation and matter. Afterward, radiation (photons) and matter (atoms) more or less went their separate ways.

Thus gravity emerges victorious in this primeval contest. With

8.2. *Cosmic Robin Hoods.*

the formation of atoms, the effectiveness of photons in pushing matter around decreases abruptly. The electromagnetic force signals to the Referee that it wants to retire from the match.

DISSIPATION AND COLLAPSE

After matter forms into atoms, these atoms attract each other gravitationally and start to lump together, as choreographed by Newton. As lumps form, more atoms rush in toward the nearest lump. As they rush in, they collide with each other, sort of like New Yorkers jostling each other as they descend into subway stations at rush hour. The colliding atoms emit photons that, since they hardly interact with atoms, escape from the mad rush, thus carrying away energy. In this way the atoms lose energy and move ever more slowly, less and less able to resist the inward pull of gravity. And thus matter collapses into more and more compact lumps. Technically, this process is known as *dissipative collapse*. The atoms dissipate their energies and collapse. (Figure 8.3.)

8.3. *An artist's concept of dissipative collapse.*

That atoms can radiate photons and dissipate energy is essential to the story. Were there no mechanism for dissipation, the energy of movement of the atoms would prevent them from collapsing into lumps. They would simply bounce off each other like perfectly elastic rubber balls. On his night out, the rake had to find ways to get rid of his excess energy before he could collapse into sleep.

Photons, playing Robin Hoods in the first act of the drama, had slowed down the concentration of matter in one region at the expense of another. Now, in the second act, in a nice plot twist, photons encourage the collapse and concentration of matter by taking away atoms' excess energy. (In this analogy, the photons starring in the first act are not the same as the photons starring in the second act. The photons that played Robin Hood are all tired out in the second act; they float about in the background. Those photons starring in the second act were emitted by the collapsing matter.)

THANKS, GRAVITY

How wonderful gravity is! Without it, we would not be. The universe would be a thinning haze without much to admire in it.

But gravity couldn't have done it alone. I stand in awe of how only the intricate intertwining of all four forces managed to bring the play off. Like the Little Red Hen, gravity asks its three friends to help. In our story, however, the cat, the duck, and the pig actually did help.

As gravity strives to bring structures out of the haze, the electromagnetic force is needed to carry the excess energy away: It's no good if everyone is zinging around. Once the particles quiet down and gravity brings the primeval hydrogen and helium nuclei face to face, the strong and the weak forces step in. The strong force causes nuclei to react with each other, thus igniting the nuclear fire that brings warmth to the vast void. The weak force is crucial lest stars become as uncivilized as nuclear bombs. Certain nuclear reactions can only proceed through the weak force. Because the weak force is, well, weak, these reactions proceed extremely slowly. As a result, the nuclear fires in stars burn at a stately pace. Meanwhile, gravity is busily collecting the ejecta left by the dying stars of the first stellar generation into planet-size bits of interstellar dirt. The electromagnetic force is keeping busy, too. It is transporting energy from the stars to warm these bits of dirt, and it is running all kinds of chemical reactions, bonding one atom to another, so more and more interesting structures can be built. It's a team effort.

This describes in broad outline the emergence of galaxies and subsequent structures. But many questions remain unanswered. The size of the initial lumps provides one focus of debate. Did matter collapse first into big lumps that subsequently broke apart into smaller lumps destined to become galaxies? Or did matter collapse into small lumps that subsequently through their gravitational attraction for each other merged into galactic-size lumps? Or did matter collapse, right from the beginning, into lumps of just the right size to become galaxies?

Astronomers generally believe that the first stars formed at the same time as galaxies. Inside a collapsing lump of matter, there are by chance here and there regions exceptionally rich in matter. These regions collapse faster and form tiny (!) lumps inside the big lumps. The rich get richer, but the superrich get richer even faster. These tiny lumps condense into stars that run through their lives quickly compared to the cosmic time scale and die explosive deaths. The explosions of these ancient stars send shock waves through the surrounding cloud of matter, thus compressing the matter and hastening its collapse. Some regions could even have been so rich in matter that they would have collapsed into primeval black holes—called primeval to distinguish them from the latter-day black holes formed from the collapse of exhausted supermassive stars. Some theorists believe that these primeval black holes could have acted as seeds for galaxy formation by gathering up matter around them. If so, then there ought to be a black hole at the center of most galaxies. Whether you subscribe to this scenario depends on what you believe the fluctuations in the density of matter looked like in the early universe. Did the density fluctuations contain lots of spikes, or were they relatively smooth? The character of the fluctuations determines how many primeval black holes were formed.

We will leave these questions of details to the experts and move on to broader issues.

NOT ENOUGH TIME TO GET RICH

The rich get richer, but there still is a problem that they often feel acutely: It takes time to become even richer.

Similarly, it takes time for the density fluctuations to grow and to collapse. Herein lies a serious difficulty for cosmologists: Was there enough time between the formation of atoms and the emergence of galaxies and stars in the cosmologically recent past?

Physicists know exactly how long ago atoms formed. How? They know how cold the universe had to be before atoms could form. They also know how fast the universe cooled. So it is simple arithmetic to figure out how much time has elapsed for the universe to have cooled to its present temperature.

Imagine coming upon a Monopoly game in progress. You notice that one player has eight thousand dollars and owns 90 percent of the hotels and houses on the board. The other players have practically nothing. You conclude that the game has been going on for quite a while. If told that the game in fact started only a few minutes ago, you can bet that the rich player must have started out with more money somehow.

Similarly, from Lifschitz's calculation, physicists know how fast mass density fluctuation grew, just as in the analogy above, you were presumed to know the dynamics of Monopoly. Since they know when atoms formed and when galaxies formed, they can conclude there just had not been enough time unless the fluctuations were large initially. In the parable, you would conclude correspondingly that the rich player must have started out with more money.

The standard rules of Monopoly mandate that each player start with the same amount of money. Instead, suppose each player starts with an amount determined by chance. The variation in the initial capital given to the players corresponds to the fluctuation in the mass density as one goes from one region of the universe to another. In the analogy you conclude that the variation in the initial capital had to be quite large. Indeed, knowing the dynamics of the game, you could estimate this initial variation.

Similarly, working backward from the time when galaxies emerged, physicists can estimate the size of the fluctuation when atoms formed. They found that the fluctuation had to be somewhere between 0.1 percent to 1 percent. That sounds tiny—surely in Monopoly it wouldn't make much difference if some players started out with 0.1 percent more money than others—but it isn't. It is too big to accord with observations.

How do we know it is too big? Because amazingly enough, astronomers can actually observe this initial fluctuation. The fluctuation imprinted itself on the photons that started to drift through the universe shortly after atoms formed.

In modern versions of the Penzias-Wilson observation of the microwave background, astronomers point their detectors in different directions in the sky. The primeval photons coming to us from different directions originated from different regions of the universe with their slightly different densities. Therefore they are expected to have slightly different characteristic energies. The difference measures the fluctuation just before atoms formed. In effect, astronomers are measuring how uniform the universe was at that time.

And the universe was darn uniform. Recent observations indicate that the fluctuation was at most a few thousandths of a percent. If so, then galaxies just couldn't have formed in the time available.

How are cosmologists going to wriggle out of this one? Tune in to future episodes.

From Hair Whorls
to the Edge of Creation

A MONKEY IN DISGUISE

When I was in the first or the second grade, we boys used to examine the backs of each other's heads. We would pretend to count the number of whorls in our friends' hair, all the while chanting, "One whorl means that you are a man; two whorls, you are a goblin; and three whorls, well, you must be a monkey in disguise!" With malicious glee, we would gang up on some poor soul and tell him that he had three whorls in his hair. The victim, in disbelief, would ask a friend to check the back of his head. At this point, the so-called friend would examine the victim's head with exaggerated care, saying, "Let's see, here is one whorl. Mmm, oh, no, here is another one! And, here is a third!" The victim, after trying in vain to examine the back of his own head in a mirror and now utterly unsure whom among his "friends" he could trust, would reluctantly go ask the teacher to examine his head.

Little did I know that our juvenile diversion represented my first encounter with that fascinating branch of modern mathematics known as topology. Certainly I had no inkling that thirty years later, physicists would use the kind of insight gained from counting hair whorls to picture what the universe might look like immediately after its creation!

THE ESSENCE OF THINGS

Topology is, in some sense, the opposite of the kind of mathematics we were taught in school. We might have learned that the surface area of a sphere is given by the radius of the sphere squared, multiplied by 4π, where π is equal to 3.14159 . . . We might even have learned the formula for the surface area of a doughnut, a formula too ugly to write down here. Some quasi-human mathematics teacher might have put us through the drudgery of answering such questions as, "What is the surface area of a sphere whose radius is 4.72 centimeters?" Topology began when some mathematician decided that he couldn't care less about what the formula for the surface area of a sphere might be. Instead, the mathematician wanted to know how the doughnut is intrinsically different from the sphere. He asked the first child he encountered and was told, "The doughnut has a hole in it!" Thus was a new branch of mathematics born.

A topologist thinks of geometric objects as being made of an extremely elastic rubber. He thinks of two objects as equivalent if, by stretching and squeezing one of the objects, he can transform it into the other object. As far as a topologist is concerned, a football and a soccer ball represent the same object. On the other hand, one cannot possibly squeeze a football into a doughnut. In short, topologists deal with the intrinsic essence of a geometric object rather than with such "accidental" features as its size. The number of holes in an object clearly provides one measure of the intrinsic essence of the object.

Consider this statement: "No matter what strategy one uses at a casino, the casino always wins in the long run." This expresses the essence of casino gambling. Similarly, topologists express one essence of dough-nuts by saying, "No matter how one squeezes the doughnut, the doughnut always has a hole."

You may feel that this is all pretty obvious, and it is. But mathematicians do not limit themselves to geometric objects in ordinary three-dimensional space, and topology can become a very difficult subject when one deals with objects in higher-dimensional mathematical space. When a mathematician describes a complicated geometric object in six-dimensional space, I find it utterly impossible to picture the object, let alone count the number of holes in the thing. The range and variety of human intelligence are truly remarkable. Just as an architect and a lawyer can be clever in different ways, the intelligence of a mathematician differs qualitatively from that of a theoretical physicist. It astonishes me that a topologist can count the holes in an object that I cannot even picture.

But what does topology have to do with physics and the real world? When I started studying physics, physicists as a group regarded topology as amusing but totally irrelevant to their field. At that time, broadly speaking, physicists emphasized specific descriptions rather than intrinsic essences. With my generation of fundamental physicists, the outlook has shifted markedly.

Perhaps my own experience is typical. My interest in topology continued into college, where the mathematician Ralph Fox introduced me to knot theory, the study of knot-tying in higher-dimensional spaces. But during one of my horrendous struggles to visualize some particularly bizarre knot, I realized my abilities lay in physics, not mathematics. I said good-bye to Professor Fox, and that was the last I heard of topology until well after graduate school. But in 1974, topology captured the imagination of fundamental physicists trying to understand Nature's basic laws. To understand why, I have to step back a bit in time.

EYE OF A HURRICANE

To study a physical system, be it the universe or a piece of metal, physicists must first describe the state or condition that the system is in. To introduce the discussion, let us first consider how metereologists describe the weather. The temperature, for example, is characterized by a number attached to every point on a weather map. Wind condition, on the other hand, has to be described by an arrow attached to every point on the map, with the length and the direction of the arrow indicating the local wind speed and the direction, respectively. The weather map represents a snapshot of the weather at a given instant.

Look at a map of the wind condition. Obviously, the meteorologist can't possibly draw an arrow for every point on the piece of paper. Instead, she shows only arrows for a number of representative points and mentally fills in the arrows for the points in between. Thus, if we see on a weather map that in New York the wind is blowing at ten miles per hour from the south and that in Philadelphia it's blowing at fifteen miles per hour from the southwest, then we figure that in central New Jersey the wind is probably blowing at twelve or thirteen miles per hour from a direction somewhere between south and southwest. In other words, in drawing the arrows, the meteorologist must obey the rule of continuity: If two points are infinitesimally close to each other, then the arrows associated with the two points can differ only infinitesimally. Weather conditions change continuously from point to point.

If we were to see on a weather map the wind conditions shown in Figure 9.1, we would know that somewhere inside the cyclone there must be a special region—the eye of the hurricane. Why? Let us enunciate explicitly what we know intuitively to be true.

Imagine filling in the circular central region of the cyclone with arrows. Remember that there is an arrow at *every* point (even though we can show only a finite number of points in any drawing) and that the arrows must be drawn respecting the rule of continuity—that is, arrows at neighboring points can differ only slightly. Well, you can't do it if you keep the lengths of the arrows fixed. As you draw more and more arrows, you will reach a point where you will not know which way to point the arrows. The requirement of continuity will impose conflicting requirements.

The situation can be avoided only if you let the arrows become shorter and shorter as you proceed inward. Eventually, at some point the arrows must vanish altogether. Mathematically, we think of the length of the arrows decreasing to zero.

Here I must emphasize the distinction between mathematical idealization and physical realization. We have discussed the mathematical problem of laying down arrows in a two-dimensional space—namely, the surface of the paper. An actual hurricane is more complicated because it exists in a three-dimensional space—air can also move up and down. At the eye, there could be an enormous updraft.

This is an example of a topological conclusion. Why topological?

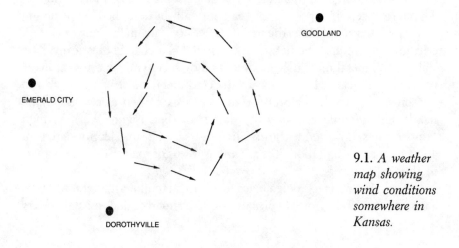

9.1. *A weather map showing wind conditions somewhere in Kansas.*

Because the hallmark of topological reasoning is evident here: No matter how we turn the arrows, the eye of the hurricane will move around, but it cannot disappear as long as we obey the rule of continuity. This is analogous to saying that no matter how we squeeze a doughnut, the hole cannot disappear as long as we don't tear the doughnut.

Remember my friend who has trouble understanding infinite space. When I got to this point and told her about topology, she kept insisting, "This doesn't make a whorl of sense! I just don't get it." Finally I interrupted her and said, "Topology is often a laughably simple subject." "Simple, eh?" she muttered. A moment or two later she exclaimed, "Oh, of course, this is so obvious. It's just that all your clumsy words were confusing me. If you turn the arrows round and round"—she waved her hands to show how—"then in the center you would go beserk, so you better stop."

VORTEX

Just like the state of the weather, the state of some physical systems can be described by numbers attached to each point in the system, while that of others must be described by arrows. (Sometimes both a number and an arrow are needed.) More interestingly, sometimes the mathematical object (number or arrow, in our example) needed to describe the state of a system can change. An exciting possibility is that the universe may be an example of such a system.

Before I go on to describe the universe, let me be more concrete. The state of certain types of metal can normally be described by specifying a number at every point in the metal. As we cool the metal below a certain critical temperature, the nature of the metal can change drastically: The state of the metal must now be described by specifying an arrow at every point in the metal. Here's why: In the cold metal, electrons pair off into twosomes. It may be helpful for you to think of two electrons linking hands and defining an arrow by doing so. (See Figure 9.2.) At high temperatures, however, the arrow disappears as the two electrons are torn apart by thermal agitation, just as various forms of matter were torn apart in the early universe.

As the metal is cooled below the critical temperature, it undergoes a sudden transition from a normal state to a superconducting state. Very

9.2. *Two electrons holding hands: a direction, indicated here by an arrow, is defined.*

roughly, in this new state an electric current can pass through the metal— via the "arrow" of the electron pairs—without any resistance.

The complete ramifications of this make for a long story that we cannot go into here. For our purposes, it is only important to know that the metal is described by numbers in the normal state and by arrows in the superconducting state.

Let us look at a thin slice of the metal. As it is cooled down, arrows suddenly appear. We may find that in some region the arrows are pointing more or less radially outward, as in Figure 9.3.

9.3. *An arrangement of arrows.*

139

You can now see why I was reminded of the hair on the back of a little boy's head. If the arrows in that figure represent the hairs, the little boys I used to associate with would exclaim that a whorl has to be somewhere under the question mark of Figure 9.3. Something must happen under that question mark.

Just as we argued when discussing the eye of the hurricane, so we conclude here that it is impossible to fill the region under the question mark by arrows of a fixed length and still obey the rule of continuity. The only way is to let the arrows shrink to zero as we proceed under the question mark. What does that mean physically? Recall that the normal state of the metal is not characterized by arrows but by numbers. Thus the metal, "knowing" that something mathematically impossible is about to happen if it insists on arrows under the question mark, saves the day by reverting back to its normal state. In other words, under the question mark is a region in which the metal is normal, surrounded by a region in which the metal is superconducting, as depicted in Figures 9.4a and 4b. Inside the normal region, electrons are forcibly divorced from their partners. Physicists call this arrangement of a normal phase surrounded by a superconducting phase a vortex.

What does topology tell physicists about the vortex? First of all, no matter how we turn and move the arrows, the vortex cannot disappear. Topology guarantees the stability of vortices.

As the arrows move and turn, the vortex moves around, much as the eye of a hurricane moves around. In the core of the vortex, electrons

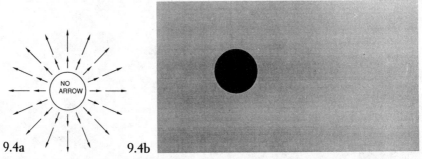

9.4a

9.4b

9.4a. *Mathematically, the arrangement of arrows shown is possible if the arrows shrink in length as one approaches the center of the vortex.* 9.4b. *The vortex is a normal region surrounded by a superconducting region.*

are forcibly divorced from their partners. Since it takes energy to tear the electrons apart, we have to put energy into the vortex. The normal core of the vortex contains energy. We can thus picture the vortex as a stable lump of energy capable of moving around. The effective size of the vortex is the size of the normal region.

I have focused on a thin slice of the metal so that I can draw the metal's state on paper. By continuity, if a given slice contains a vortex, then a neighboring slice must also contain a vortex. Thus, as illustrated in Figure 9.5, superconducting metals actually contain vortex *lines*, in a pattern reminiscent of tornados snaking across the metal.

In summary, topology guarantees the existence and stability of the vortex. That the state is described by arrows is essential for topological existence. Given the temperature distribution shown in Figure 9.6, we can't deduce anything about what the temperature might be at the point indicated by the question mark. It could be 50°, 60°, or 70° for all we know. Distributions of arrows such as those shown in Figures 9.1 or 9.3, however, allow us to deduce that "something" must appear at the center.

In the physical world, permanence and stability are often lacking. A cotton candy cloud drifting lazily in the sky could eventually disappear as the water molecules in it hop off. Even an atom can be torn apart and be no more: The electrons in it can go elsewhere. But try as you may, turning the arrows this way or that, you can't make the vortex disappear, in the same way that by stretching without tearing you can't make the

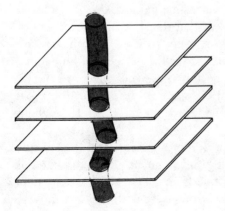

9.5. *The vortex is actually a vortex line snaking across the metal, much like a tornado.*

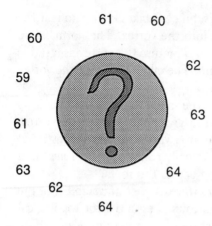

9.6. *A distribution of temperature on a weather map. In contrast to the situation in Figure 9.3, here we cannot state with certainty what the temperature distribution has to be under the question mark. It appears likely that the temperature there would be in the low 60s, but it could be in the 40s or the 80s. Indeed, it could be just about anything.*

hole in a doughnut disappear and turn the doughnut into a ball. Topological objects endure.

HEDGEHOGS IN THE EARLY UNIVERSE

Now that we have seen how topological concepts can enter into physics, we are ready to examine the early universe.

A revolution occurred in physics around the year 1971: The very way in which physicists describe Nature changed. A detailed account of the revolution would occupy volumes. Here we focus on one particular aspect of this revolution. Previously, physicists had thought that the state of the world could be described by specifying a number at every point in space. But around 1971, it became clear that the state of the world must be described by specifying an arrow at every point. (These arrows are associated with a force field needed to describe the behavior of fundamental particles. The precise nature of these arrows does not concern us here.) Physicists believe that the arrows that characterize the state of the world are all lined up, pointing in more or less the same direction, throughout the universe.

In 1974, the Dutch physicist Gerhard 't Hooft and the Russian physicist Aleksandr Polyakov, working independently, raised what was regarded at that time as a curious question. They asked what would happen if in some region of space the arrows were not lined up but were instead arranged in a three-dimensional version of Figure 9.3. (In other words, we have arrows pointing radially outward, much as the pins on a spherical pincushion. To visualize this, you can also think of a hedgehog or a porcupine. When a hedgehog is threatened, it curls itself up into a more or less spherical ball with all its sharp spines pointing radially outward. The arrows correspond to the spines.) The fact that this question had never even occurred to most fundamental physicists indicates how peripheral a niche topology occupied in their collective consciousness.

But having absorbed the preceding discussion on topology and cold metal, you can see immediately the answer to the question of 't Hooft and Polyakov: Some kind of vortex carrying a certain amount of energy can exist. Indeed, Polyakov was inspired to name this physical entity the *hedgehog*. Inside the hedgehog, just as in the superconductor vortex, the underlying physics averts a crisis by creating a region inside which the arrows simply shrink to zero. Just like the vortex, the hedgehog is a compact bundle of energy guaranteed by topology to stay together as an entity.

The grand unified theory mentioned in Chapter 7 contains just the kind of arrow with which hedgehogs can be made. Thus, physicists who believe in the grand unified theory also believe in hedgehogs. The size and mass of the hedgehog are determined by the characteristic length and mass scales of the grand unified theory: The hedgehog turns out to be extremely tiny, about 10^{-16} times smaller than the proton, and extremely massive, about 10^{17} times more massive than the proton.

The notion that a physical entity could be constructed topologically was a stunning revelation to fundamental physicists, most of whom were unfamiliar with vortices in superconductors. (I for one was never taught anything about vortices—such is the effectiveness of education.) But a surprise was yet to come. Polyakov and 't Hooft went on to study the hedgehog's physical properties. Remarkably, the theory indicates that a magnetic force field emanates from the hedgehog. The hedgehog is the elusive and legendary magnetic pole!

The magnetic pole (also known as the magnetic monopole) is the unicorn of physics history, long sought after but never seen. Electromagnetic theory displays a curious lack of symmetry between the electric and the magnetic. Nature is full of electric force fields emanating from electric charges. But nobody has ever seen a magnetic charge or pole. In high school we learned that a magnet always comes with two ends or poles, called by tradition the north pole and the south pole. If we try to cut a magnet into two halves, we end up with two magnets, each with two poles. (Figures 9.7a and 7b.)

In 1931 the English physicist Paul Adrien Maurice Dirac speculated that magnetic poles exist and worked out their properties. But theorists were never able to fit the magnetic pole into an acceptable theory. So it was with great excitement that physicists greeted the news that the magnetic pole had been constructed as a topological entity. I remember arriving at CERN, the European Center for Nuclear Research in Geneva, Switzerland, for a visit in the summer of 1974. Immediately upon my arrival, one of my colleagues, flushed with excitement, showed me a copy of 't Hooft's paper. The topological construction of the hedgehog, at once simple and profound, bowled me over right away.

Given that a magnetic pole has never been seen, experimenters immediately asked whether they could make one in the laboratory. (After all, some circus operator tried recently to make a unicorn, amid outcries from animal-lovers.) But unfortunately, the theory indicates that the amount of energy packed inside the hedgehog is much larger than the energy our particle accelerators can possibly deliver.

As in all good mysteries, the mystery of the missing hedgehog took on an interesting plot twist. Around 1978, the Russian physicists Y. B. Zel'dovich and M. Y. Khlopov and, independently, the

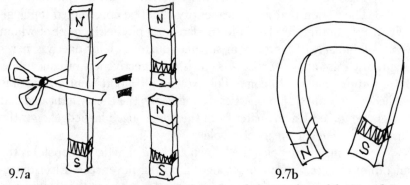

9.7a   9.7b

A bar magnet has a north end and a south end, here indicated by N and S.
9.7a. If you cut a bar magnet in two, you get two bar magnets. You don't get two chunks of iron, one with only a north end, the other with only a south end. 9.7b. A horseshoe magnet is a bent bar magnet.

American physicist John Preskill pointed out that hedgehogs should have been created in abundance in the early universe. To understand this point, we have to picture what the universe looked like moments after Creation.

WORLDS OUT OF TOUCH

First there was a big bang, the exact nature of which lies beyond contemporary physics. The force of the explosion was such that the universe expanded outward at a tremendous speed. Over the eons, the expansion of the universe gradually slowed (as we saw in Chapter 4); the expansion rate now observed by astronomers, rapid though it is in human terms, is but a pale shadow of the rate right after Creation.

With the universe expanding so rapidly, different regions of the universe cannot possibly keep in touch with each other. According to Einstein, the fastest signal one can send is a light pulse. But we are talking about literally moments after Creation—hedgehogs were actually created 10^{-36} seconds or so after the bang. To make the point here, we might as well talk of the somewhat more comprehensible time of a billionth of a second. In a billionth of a second, even light traverses only 30 centimeters. Thus, a billionth of a second after the big bang, two regions of the universe more than 30 centimeters apart are not in touch. When you "reach out and touch someone," you have to allow enough time for the signal to get through. Therefore, the early universe was divided into numerous worlds totally out of touch with each other. Meanwhile, with the universe expanding furiously, the different worlds were moving farther apart.

It is amusing to construct a hypothetical example of how hedgehogs or vortices could appear in cultural anthropology. Picture a continent populated by many civilizations, each in contact with only its neighbors. Suppose that in the easternmost civilization people always pray facing east, that in the westernmost civilization people always pray facing west, and so forth. In fact, suppose that as a cultural anthropologist travels along the coast, he or she finds that in each of the civilizations encountered people always pray facing the sea, so that the direction of worship rotates by 360° as one makes a complete circuit around the continent. The situation is then precisely as depicted in Figure 9.3 if the arrows are taken to represent the direction of worship. Now suppose the anthropologist ventures inland. Assume that the rule of cultural continuity holds so that as the anthropologist travels, he or she always finds the direction of worship changing slowly. But then something has to give. Somewhere in the heart of the continent there would be a civilization so confused about the direction in which they should face that they might have given up and become atheists. The pocket of atheism is the vortex or hedgehog.

In the early universe different regions were in contact only with neighboring regions. Imagine a group of nonoverlapping spheres. In each of these spheres are arrows describing the state of the universe within. We expect the directions of the arrows in two nonneighboring spheres to be completely independent of each other. Fill in the spaces between the spheres with arrows according to the rule of continuity. Clearly, in many regions the arrows would not match, and hedgehogs would have to be formed. See Figure 9.8. The same story occurs in a metal as it cools into a superconducting phase. The arrows in different regions are pointing every which way, thus forcing vortices to form.

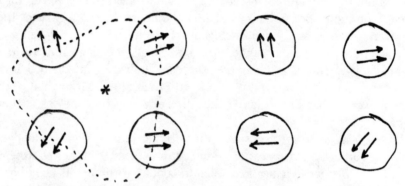

9.8. *Arrows describing the state of the world in the early universe. An imaginary being traveling along the dotted line will find the arrows turning by 360° as he completes the circuit. Just as the cultural anthropologist in our analogy concluded that a pocket of atheism must exist in the heart of the continent, our imaginary being could conclude that a hedgehog must lurk somewhere inside the dotted curve, perhaps at the place marked with a star.*

Reasoning along this line, Zel'dovich, Khlopov, and Preskill easily estimated by using the statistical law of chance the number of hedgehogs per unit volume of space produced in the early universe. Knowing how much the universe has expanded since then, they can therefore determine how many hedgehogs, or magnetic poles, ought to be floating around today in a given volume of space. The reasoning here depends on the fact, guaranteed by topology, that a hedgehog, once produced, cannot simply disappear.

This simple calculation shows that the number of magnetic poles now floating about is enormous. We should be able to find a magnetic pole in every closet, in total contradiction to observation.

As I have introduced quite a few concepts, of which some were needed for illustrative purposes only, it may be good to summarize the revelant points thus far: (1) Topology guarantees the permanence of vortices and hedgehogs once they are created. (2) Particle physicists discovered in 1971 that the state of the universe should be described by arrows at each point in space rather than by numbers. (3) The early universe was expanding so rapidly that different regions were out of touch and thus the arrows describing the state of the universe in one region could point in a totally different direction from those in another region. (4) Topology guarantees that this misalignment of arrows necessarily implies the presence of hedgehogs. (5) Since the hedgehogs—which turn out to have properties of magnetic poles—cannot disappear, they should still be around. (6) Where are they?

MYSTERY OF THE MISSING HEDGEHOGS

The mystery deepens. The skeletons are not in the closets and are nowhere to be found. Where did they go?

A rather nifty resolution of the mystery was invented in 1980 by Alan Guth, who showed that the expansion of the universe could have gone into an "inflationary era" after the hedgehogs were created. In this era, the universe doubled its size every 10^{-36} seconds or so. Thus, after ten periods of 10^{-36} seconds or 10^{-35} seconds, the universe had expanded to 1,024 times its former size. After one hundred periods of 10^{-36} seconds or 10^{-34} seconds, the universe had expanded by a huge factor of about 10^{30}. This type of expansion is so astonishingly rapid that Guth reached into the American economic experience of the late 1970s to describe it. Anyone who has lived through runaway inflation or struggled to pay off a high-interest loan would appreciate how fast the universe is supposed to have been expanding. We are talking about 100 percent inflation, or interest every 10^{-36} seconds!

According to the scenario, our visible universe with its almost inconceivable vastness was inflated from a tiny region, a region so tiny that it could not have contained more than one hedgehog or so. Thus the inflationary scenario resolves the mystery of the missing hedgehogs: There are at most only a couple of hedgehogs within our visible universe.

The mystery of the missing skeleton is resolved: The skeleton is still there, but the closet has expanded a zillion times since the crime was committed! Were this a crime mystery, the reader would surely fling the book out the window in disgust. But because the inflationary scenario not only explains the missing hedgehogs but also solves some other outstanding cosmological puzzles, many theorists have embraced Guth's idea with great enthusiasm.

In contrast to the theorists, experimenters searching for magnetic poles were bitterly disappointed by the inflationary-universe scenario. However, the spirit of empirical science is such that experimenters have stubbornly persisted, in the hope of finding a magnetic pole and thus proving the inflationary theorists wrong. At present, the situation is excitingly unclear. On February 14, 1982, shortly before 2:00 P.M., something resembling a magnetic pole passed through a detector rigged up by Blas Cabreras at Stanford University. The surprising discovery hit front pages, but sad to relate, that singular event never recurred. Other experimenters had all failed to see the magnetic pole. The growing consensus is that the sighting was due to an unexplained glitch in Cabreras's detector. The probability that the one magnetic pole in the entire visible universe just happened to be loitering around Palo Alto on Valentine's Day in 1982 is, of course, practically nil.

Physicists, like a mystery reader who has thought of a solution halfway through the book, have the uneasy feeling, even as they congratulate themselves for thinking of a rather nifty solution, that the Ultimate Mystery Writer may have an even niftier solution up His sleeve. Furthermore, over the past few years theorists have found that some of the consequences of the inflationary scenario do not check out in detail against the available evidence. There may well be other plot twists down the line! Even if the inflationary scenario proves to be correct, there appear to be quite a few details we do not yet understand.

The inflationary scenario has become tremendously popular with theorists eager to hide various unobserved entities. My colleagues and I might sit around discussing some theoretical idea. "Oops, we've got way too many of these X particles!" someone might interject. Holding off panic, these days I might just reply, "Yeah, no sweat, let's just get some inflation going after the X particles were created." The trick is not to get the goodies we do want, such as protons and electrons, inflated away as well.

THE INFLATIONARY UNIVERSE

The inflationary universe is perhaps the most exciting idea introduced into cosmology in recent years. What caused the universe to inflate?

To melt ice into water we have to put in a certain amount of energy in the form of heat. Conversely, when water turns into ice it gives out energy. (A refrigerator is a machine designed to take this energy away and dump it, ultimately, into the ambient air.) The transformation of water into ice is another example of what physicists call a *phase transition*. Water and ice are both made of the same water molecules but are arranged in different ways.

As the early universe cooled, it, too, could have undergone a phase transition analogous to the transition from water to ice. (Instead of a rearrangement of water molecules, a rearrangement of a fundamental force field governing the behavior of various subnuclear particles would have been involved.) Just as in the transition from water to ice, a large amount of energy is dumped into the universe, and hence the universe goes into a wild expansion.

As I mentioned, many cosmologists fell in love with the inflationary universe because it solved, besides the mystery of the missing hedgehogs, a number of other cosmological puzzles.

Consider the homogeneity of the universe, for instance. On small scales, the universe is certainly not homogeneous. As realtors like to say, the three most important words in real estate are neighborhood, neighborhood, and neighborhood. We seem to be in a cozy neighborhood, with a steadily burning star. Another neighborhood is just deserted—one vast, empty space. However, when cosmologists look at the universe on large enough scales and count up the number of galaxies per unit volume, they find the number density to be more or less the same wherever they look. This is another manifestation of the notion that no one place is more special than another in the universe.

This observed homogeneity vexed cosmologists for a long time. For a period after the big bang, different regions of the universe were not in touch, as I mentioned earlier in this chapter. How could the universe have smoothed itself out? The number density of galaxies in different parts of the universe ought to be different.

Consider two points in the universe a zillion light-years apart now. How close were they to each other one second after the big bang? Without inflation, they would have been in two regions that haven't yet communicated with each other. With inflation, however, the two points were much closer together and could have been in the same region. The inflationary scenario solves the homogeneity puzzle simply by saying that runaway inflation had stretched a smallish region in the early universe into the enormous portion of the universe visible to us now.

If so, then we have seen only a small portion of the universe. If we can look at the universe at even larger scales, inhomogeneity may show up.

The inflationary scenario has also run into some severe difficulties. Economists can appreciate one of these, namely the difficulty of a graceful exit. How does one bring inflation to an end without causing a recession? Physicists have to invent various schemes to ease the universe from inflation back into steady expansion. More on inflation later.

COSMIC STRINGS

Topological considerations are important not only for our understanding of the extremely early universe but also for our understanding of one main event of the universe's later evolution: the emergence of galaxies. As we saw in Chapter 8, Newton had already remarked on the inherent instability of gravity. A region richer in matter than the surrounding regions would become even richer by pulling in matter from the surrounding regions. In this way, galaxies were born.

Everyone in the galaxy-making business agrees that Newton was right. Given an initial irregularity in the matter distribution, or in technical terms, a primordial mass density fluctuation, the irregularity will grow. But where does the primordial mass density fluctuation come from?

This question has long troubled physicists. Few if any can accept the notion of Divine Interference forty thousand years after the bang. The primordial fluctuation had to be just right for galaxies, stars, and the rest to form at the time they actually did form. (We know, from the calculations of Lifschitz and others, the growth rate of the fluctuation, so we can extrapolate backward to determine how large the primordial fluctuation had to be.)

Physicists have racked their brains for an answer to this question. In the past few years, one particularly fascinating suggestion has attracted an increasing number of adherents: the suggestion that cosmic strings triggered the primordial fluctuation. Having carefully laid the groundwork, I can now explain easily what a cosmic string is.

Go back to the vortex lines found in cold superconducting metals. Recall that the arrows formed by the pairs of electrons may be arranged topologically to give a vortex line snaking across the metal. But as far as the topology is concerned, it does not matter whether the arrows correspond to hairs on a boy's head, pairs of electrons, or the arrows describing the state of the universe at each point in space. In 1976, the English physicist Tom Kibble suggested that perhaps there are vortex lines snaking across the cosmos itself. Just as vortex lines could form as a piece of metal cooled, so they could form on the cosmic scale as the universe cooled.

If true, this suggestion would drastically revise our picture of the universe: At one point the sky may have been crisscrossed with what physicists call cosmic strings. (See Figure 9.9.) These strings are extremely

9.9. *An astronomer contemplating a heaven of fettuccine?*

thin but staggeringly massive: Theorists estimate that cosmic strings have a thickness of about 10^{-29} centimeter and a mass of about 10^{19} kilograms per meter! Again, the thickness and mass of the string are determined by the characteristic length and energy scales of the grand unified theory. To get a feel of how massive these strings are, consider that a 100-meter segment of string weighs as much as the moon.

ORIGIN OF FLUCTUATIONS

In 1980, the aforementioned Y. B. Zel'dovich suggested that cosmic strings may have generated the primordial density fluctuations. Some cosmic strings might have born snaking across the entire universe. Others might have formed closed loops. Since their births were due to the haphazard laying down of arrows, they would all be twisted and tangled. The natural elasticity of these strings caused them to writhe and vibrate, thus radiating away energy. Loops of cosmic strings would shrink as they lost energy. Strings would also crash into each other.

By their enormous heft, cosmic strings would pull the ambient matter toward them gravitationally. Thus, all this rock and roll and twist and shout by cosmic strings could have produced the irregularities in the distribution of matter needed for the formation of galaxies and the emergence of structures. Loops of cosmic strings could have acted as seeds for galaxies as matter fell in toward them.

In recent years, a number of physicists, notably the Russian-American Alex Vilenkin, the Englishman Neil Turok, the Pakistani-American Qaisar Shafi, the Greek George Lazarides, and many others have been working hard to determine whether the string mechanism of galaxy formation agrees in detail with the observed distribution of galaxies.

COSMIC LENS

How might we actually see cosmic strings? Suppose there is a cosmic string between us and a distant galaxy. Because of the enormous mass contained in the string, light emitted from the galaxy is bent toward the string as it passes by the string. (Remember how the old man's toy implied that gravity bends light?) In other words, the cosmic string acts as a cosmic lens. We are fooled into thinking that we are seeing two galaxies, but on closer inspection the two galaxies turn out to be identical. Thus, if astronomers start to see double, they may be discovering cosmic strings. (See Figure 9.10.)

NEW FORMS OF MATTER

Topology has brought us new forms of matter. Previously, physicists thought of all matter as made up of fundamental particles such as the proton and the electron. The recent developments described here taught us that new forms of matter may be made by "twisting" the force fields of particle physics. We are seeing how these new forms of matter may affect the dynamics of the universe.

I digress for a moment to describe a technical point that, if left unmentioned, might confuse some readers. To describe the state of the universe, more than one type of arrow may be required. Whether a given type of arrow arranged topologically makes a hedgehog or a vortex line depends on the type of arrow and on the details of the theory. (Theorists have scrambled to learn topology to answer precisely this sort of question.) Most theorists believe in the type of arrow that makes hedgehogs. Fewer theorists believe that the type of arrow that makes vortex lines is also present. The consensus is that cosmic strings are less likely to exist than

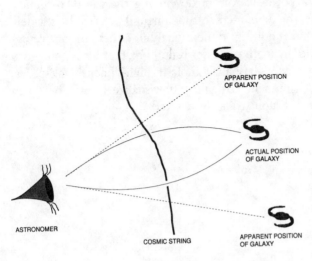

9.10. *The light coming from a distant galaxy is bent by a cosmic string toward an astronomer. The light passing by the string on its two sides is bent in two opposite directions. The astronomer, seeing the light coming in two different directions, thinks that she is seeing two galaxies.*

153

hedgehogs, just as in certain circles, griffins are considered more mythical than unicorns.

The reason why theorists working in this business sometimes sound like medieval theologians is clear: It is hard to go out and check on the existence of these cosmic entities. How we envy our colleagues who study metals! They can actually photograph the vortex lines in a piece of cold metal.

METHODOLOGICAL UNITY

From hair whorls to cold metals, from magnetic poles to cosmic strings, topology provides the common thread, underlying the methodological unity of physics, a unity that was far from apparent at one time. In graduate school they taught me neither the physics of cold metals nor the dynamics of galaxy formation. What did galaxies have to do with the itsy-bitsy particles fundamental physicists were supposed to study? One of the most stirring stories in contemporary physics, the linkage between the very large and the very small, was totally unexpected. Physics has emerged healthier, with cross-fertilization between various apparently unrelated areas.

By necessity, the examples I have given do not reflect the full power and glory of topology. Nor have I listed the full spectrum of topological applications in fundamental physics. But let me mention that topology is playing a powerful role in developing the so-called higher-dimensional theories, the *dernier cri* in fundamental physics. It is widely believed that the world is not four-dimensional but either ten- or twenty-six-dimensional. I told my math professor Ralph Fox that I was going into physics since I can't possibly visualize knots in high-dimensional spaces. Twenty years later, I am now struggling to visualize bizarre beasts in ten-dimensional space. Ralph would have liked that.

10

Ghost Riders in the Sky

INVISIBLE STUFF

In the title song from the movie *Ghostbusters,* Ray Parker, Jr., sang of finding an invisible man sleeping in your bed. Imagine coming home after a tiring day. With a sigh of relief, you flop into bed. But a creepy feeling immediately seizes you—there is somebody else in the bed. You jump up and look around. You see nothing, but from the way the bed sags, you know whoever it is weighs a whole lot. You grope around with your hand, but you feel only thin air. What ya gonna do? Rush out of the room screaming? Who ya gonna call?

Think of our galaxy as our collective bed. The galaxy contains some 10^{11} stars, of which many have planets orbiting around them. We happen to inhabit the third planet of a rather unremarkable star near the edge of the galaxy. Perhaps we share the galaxy with a multitude of other creatures.

During the past decade or so, observational astronomers have slowly come to the disturbing conclusion that there is something else—a mysteriously invisible stuff—pervading our galaxy. The invisible stuff, whatever it is, far outweighs all the stars we can see in our galaxy. Astronomers and physicists have reacted to the news with consternation. There is no one to call.

THE WAY THE GALAXY SAGS

Astronomers deduce the presence of the invisible stuff more or less in the same way that the person in the song knew about the invisible man—by the bed sagging. In this case, the bed is the collection of stars that make up the galaxy, and the sagging shows up in the way the stars move.

Every child knows that on a merry-go-round you've got to hang on or the centrifugal force would fling you outward. Similarly, as the galaxy rotates, an individual star is kept on its circular path by the inward gravitational pull exerted by the galaxy.

Sir Isaac Newton taught us that the faster a star goes around, the stronger the gravitational pull on that star has to be. In playgrounds, kids squeal with delight as they ask their fathers to spin the merry-go-round faster and faster. "Okay, I will go faster, but you have to hold on tighter!" Astronomers utilize this same principle. By painstakingly measuring how fast a star goes around, they can determine the gravitational pull on that star.

Newton also told us the strength of the gravitational force between any two objects. Thus we can add up the gravitational pull on any given star by the zillions of other stars in the galaxy. The big surprise was that the inward gravitational pull thus added up is not enough to account for the pull deduced from the motion of the stars. The stars are moving too fast! There must be something else, somehow undetected by astronomers, pulling the stars inward. Kind of strange, isn't it?

CONSERVATIVE RADICALS

Hey, wait a minute! you may object. How do you know that the gravitational pull between two stars isn't stronger than what Sir Isaac said? After all, Newton's law of gravity has been tested accurately only over distances the size of the solar system by studying the motion of planets and, in recent years, of spacecraft. And here we are talking about the immense distances between stars. Yes, indeed, that is logically possible. But the point is that Newton's law of gravity has been fitted seamlessly into Einstein's theory of gravity, which, aside from its tremendous theoretical elegance, has been tested to a fair degree of accuracy. One cannot arbitrarily modify Newton's law without bringing the whole edifice down.

Scientists are characteristically conservative radicals. When faced with a surprise, they tend to adopt the most conservative of the various radical options capable of explaining the observation. In this case, the overwhelming majority of physicists find it far easier to believe that something mysterious and invisible is out there than to jettison what we know about gravity.

Since the physics involved—namely, Newton's laws of motion and gravity—is extremely well established, and since no one has credibly disputed the astronomical measurements, most physicists believe in the existence of invisible stuff, referred to in the technical literature as *dark matter*. But what is it? No one knows for sure.

ALL AROUND US

One important clue comes from the distribution of the dark matter. Is it concentrated in a dense lump in the center of the galaxy, or it it diffuse, pervading the entire galaxy?

We can determine the distribution of the dark matter by comparing the motion of two stars, one closer to the center of the galaxy, call it star B, and the other, call it star A, farther out, near the edge of the galaxy. As a simplifying assumption, we may take the galaxy to be spherical.

Suppose the dark matter is concentrated in the center of the galaxy. Then the same amount of dark matter would be pulling on the star farther out as on the star closer in. (See Figure 10.1a.) (The force on the star farther out would be smaller than the force on the star closer in, of course.)

In contrast, suppose the dark matter pervades the universe. Then

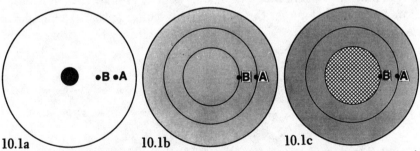

10.1a   10.1b   10.1c

10.1a. *If the mysterious dark matter is concentrated in a lump at the center of the galaxy, the same amount of dark matter would be pulling star B closer in than star A, farther out.* **10.1b.** *If the dark matter pervades the galaxy, then less dark matter would be pulling on star B than on star A. (Remember figure P.7?)* **10.1c.** *This picture elucidates the situation in* **10.1b** *further. Only the matter contained within the cross-hatched region pulls on star B. All three Figures are two-dimensional representations of a three-dimensional situation.*

the two stars are moving inside a distribution of dark matter. (See Figure 10.1b.) How do we figure out the gravitational pull of the dark matter on them?

Remember at the beginning of the book all the talk about the underground gravity express? In the Prologue, I discussed Newton's theorem that if you are inside a spherical distribution of matter, you would feel the gravitational pull of only the stuff closer to the center of that distribution than you are. In the present context, that means more dark matter would be pulling on star A, the one near the edge of the galaxy, than on star B, the one closer to the center. Only the stuff indicated by cross-hatching in Figure 10.1c would pull on star B.

That's it! The relative motion of the two stars would be different according to whether the dark matter is concentrated or diffuse. By measuring the motion of stars, we can determine how the dark matter is distributed. We spoke of comparing the motion of a pair of stars, but since there are zillions of stars in each galaxy, we can look at numerous pairs and thus map out the distribution of the dark matter fairly reliably.

There is nothing particularly mysterious or profound about this discussion. Is the invisible stuff on your bed spread out all over, or is it in a lump at the center? You can tell by the way the bed sags.

Well, astronomers have spent countless lonely nights measuring the motion of stars. Their data show clearly that the dark matter is more or less uniformly distributed throughout each galaxy rather than concentrated in a single lump at the center. Dark matter appears to fill up the space between the stars. Whatever it is, the stuff may be floating around you and me!

FLOTSAM IN THE COSMIC SEA

There is a lot of dark matter. Stellar movements indicate that the dark matter in the galaxy outweighs the total collection of stars in the galaxy. This astonishing conclusion completely revises our picture of galaxies. The stars are now seen as insignificant bits of flotsam bobbing about in a sea of dark matter.

Astronomers have gone farther and studied the movements of

galaxies themselves. In the immense depths of space, there are pairs of galaxies married to each other through their mutual gravitational attraction. Indeed, galaxies tend to occur in clusters. Our home galaxy, the Milky Way, is grouped with about eighteen other galaxies to form a cluster known as the Local Group. Some forty-five million light-years away is another cluster, the Virgo Cluster, containing hundreds of galaxies. Clusters of galaxies are in turn grouped into enormous superclusters, some as large as several hundreds of light-years in size. For instance, our Local Group and the Virgo Cluster are two members of a supercluster known simply as the Local Supercluster.

By studying the motion of galaxies in clusters and the motion of clusters in superclusters and following the line of reasoning outlined above, astronomers concluded again that the entire universe is pervaded with dark matter. The gravitational pull on galaxies can be accounted for only if the space between galaxies is also uniformly filled with dark matter.

We are therefore led to a revolutionary picture of a universe filled with dark matter within which concentrated lumps of ordinary visible matter—the galaxies, the stars, and the planets—move. The total mass of dark matter is about ten times the total mass of visible matter. However, since the dark matter is more or less uniformly distributed throughout the universe and since the distances between stars, not to mention the distances between galaxies, are so inconceivably vast, the mass density of dark matter is much lower than the mass density of visible matter. There is only about 1 gram of dark matter in 10^{30} cubic centimeters. Imagine a cube 100,000 kilometers on the side. That cube would contain on the average 1 gram of dark matter. The dark matter is extremely diffuse.

(To a first approximation, dark matter is distributed uniformly throughout the universe, but from our understanding of gravity, we expect the dark matter to be somewhat more concentrated in and around galaxies and stars. Just as the dark matter tugs gravitationally at galaxies and stars, so galaxies and stars also pull the surrounding dark matter toward them.)

WHAT IS IT?

For the past several years, various theorists have been busily proposing candidates for the dark matter. My friends and I all have our favorites. The most mundane and conservative view (but also, as we will see, an essentially untenable view) holds that the dark matter is just cosmic debris, made up of the same stuff as ordinary visible matter—protons, electrons, and neutrons.

Of course, when astronomers speak of mysteriously invisible stuff out there, they mean that they are unable to see it with their telescopes. Besides optical telescopes, which are designed to detect light, modern astronomers have at their disposal a whole array of other types of telescopes, designed to detect other forms of electromagnetic emissions, such as gamma rays, X rays, ultraviolet rays, infrared rays, radio waves, and microwaves. They have therefore concluded that the dark matter must be such that it does not emit any of these electromagnetic waves. But since astronomers cannot shine a giant flashlight into space, it is possible that the dark matter consists of just ordinary rocks or planets floating out there. While these rocks or planets can reflect some of the light reaching them from distant stars, they are too far out in interstellar space for us to see them.

ORDINARY IT ISN'T

The trouble with this suggestion is that we are at a loss to explain how all these rocks and planets got there in the first place. How many such hypothetical planets do we need to account for the dark matter? We would need an awful lot, because planets are featherweights in the cosmic scheme of things. Why can't we imagine having enormously massive planets? We cannot, for a simple physical reason. The core of a planet, squeezed by the inward gravitational crush of the planet's outer layers, gets hot. We all know about volcanoes and hot springs on our home planet. The larger the planet, the stronger is the gravitational squeeze, and the hotter the core gets. The core temperature of a planet Jupiter's size is almost high enough to ignite nuclear burning. Thus, anything more massive than Jupiter is no longer a planet, but a star, and hence would have been seen. Think of planets as Hollywood starlets that are trying to catch fire but can't quite. Invisible planets would have to be tiny, and hence we would need an awful lot of them.

Our present understanding is that planets are formed from matter ejected in violent explosions of ancient stars. The general scenario of stellar birth and death has been worked out in fair detail, and it is impossible to see how interstellar and intergalactic space could come to be packed with such a large number of unseen planets or rocks.

But suppose that the dark matter consists of, instead of planet-sized lumps, a very fine dust or a gas. This suggestion is somewhat more plausible because gas clouds and dust clouds are known to exist in interstellar space. But were the universe permeated with dust in sufficient quantity to account for the dark matter, distant astrophysical objects would be obscured. Thus, observations indicate that there cannot be too much gas or dust around.

You raise your hand. What about black holes? Everyone knows that black holes, formed either from the collapse of extremely massive stars, or in the primeval universe, are so massive that even light is gravitationally sucked in. But while black holes are themselves, well, black, they would betray their presence in the process of sucking in the ambient interstellar gas. The gas, while rushing toward the hole of no return, becomes so agitated that it emits intense bursts of radiation. Indeed, this is how astronomers detect candidates for black holes.

The nail in the coffin burying the suggestion that the dark matter is just ordinary matter that just happens not to be glowing comes from another direction. Recall that by studying the composition of stars and interstellar gas clouds, astronomers have determined the relative abundances of hydrogen, deuterium, helium, lithium, and other elements in the universe. From this information, and using the theory of nucleosynthesis as outlined in Chapter 6, physicists can deduce with a fair amount of confidence how many protons and neutrons there are in the universe.

The number deduced in this way turns out to be just enough to make all the observed stars. Many physicists accept this argument as conclusive proof that the dark matter cannot consist of ordinary matter, of the kind of stuff you and I and stars and galaxies are made of. As in a detective mystery, the most obvious suspect has been cleared.

So, what is it?

INVENT ONE TO FIT

At this point, particle physicists came barging into the dark matter puzzle. "If it ain't the garden-variety stuff, then it can only consist of one of those weird particles that we love so much, and if it ain't one that we know, we can try to invent one to fit!"

What characteristic properties must the candidate particle have? First of all, it must be stable over cosmologically long times. This requirement rules out just about every one of those particles produced in accelerators over the past forty years or so: They all disintegrate in a zillionth of a second. Secondly, one must produce a scenario whereby the universe came to be suffused with these particles. Thirdly, the particle must interact very feebly, if at all, with ordinary matter. We are floating around in a sea of these particles, and we do not notice them at all. These particles are like ghostly phantoms passing right through you and me.

LE PASSE-MURAILLE

The French conteur Marcel Aymé told a marvelous story about an otherwise unremarkable bachelor named Dutilleul who had the singular ability of passing through walls. According to Aymé, this ability was caused by the "helicoidal hardening of the strangulatory wall of the thyroid vesicle." The story went on to recount the hilarious misadventures of Dutilleul, including the obligatory encounter with a blonde. Aymé explained that Dutilleul's sartorial flair made him irresistible, observing that "Nothing speaks more to the imagination of young women today than knickerbockers and a pair of horn-rimmed glasses. It smacks of the movie director and makes you dream of cocktails and California nights." You can imagine how this encounter inevitably doomed Dutilleul to a tragic end. Regrettably, it would really take us too far afield to go into that. But I do want to talk about passing through walls.

Physicists are generally reputed to be party poopers and killjoys, so the reader probably is expecting me to say that passing through walls is impossible. Yes, according to the laws of physics it is impossible for a human to pass through walls. Indeed, I will stake my professional reputation on that assertion. Besides, I checked with my family doctor, and he has never heard of thyroid vesicles.

BRISTLING WITH FORCE FIELDS

More seriously, let us examine more closely what happens when we push one solid against another. The nuclei of the atoms making up a solid are linked together into a lattice, over which the electrons swarm. The space inside a solid is actually bristling with electromagnetic force fields. Any intruding atom will be decisively repelled.

Indeed, any particle that feels the electromagnetic force cannot get through a wall without doing damage to the wall. Of course, a particle endowed with enough energy can bull its way through by drilling a hole through the wall.

When a particle affects the motion of another particle, physicists say the two particles interact with each other. If we want to send something through a solid wall, that something cannot participate in the electromagnetic interaction. But what about the other interactions?

THE FOUR INTERACTIONS

As I mentioned in Chapter 7, physicists know of four fundamental interactions between particles—strong, electromagnetic, weak, and gravitational interactions. As the names strong and weak suggest, the strong interaction is much stronger than the electromagnetic interaction, the weak interaction much weaker. But as we saw in Chapter 3, the gravitational interaction is the most feeble by an enormous margin.

The reason that we are aware of only the electromagnetic and gravitational interactions in our daily lives is because they both reach over long distances. As we saw in the Prologue, one of Newton's great insights was his realization that gravity has a long range. In contrast, the strong attraction between two protons decreases precipitously as the separation between them increases. Two protons attract each other only when they are practically next to each other. The strong interaction is said to have a short range. The range of the weak interaction is a thousand times shorter still.

According to the equivalence principle, all particles interact gravitationally. Remember from Chapter 1 that when we drop something we can say equivalently that the floor is rushing up? Gravity is universal. While all particles interact gravitationally, they do not necessarily participate in all four interactions. Of the three most familiar particles—the electron, the proton, and the neutron—the electron does not interact strongly, while the proton and the neutron participate in all four interactions. (In fact, a particle that interacts strongly will participate in all four interactions.)

The difference in range and strength of the four interactions, together with the fact that they affect each particle selectively, accounts for the rich diversity of physical phenomena. For instance, everyday objects are large because atoms are very large on the scale of fundamental particles. If we imagine magnifying the hydrogen atom until its nucleus, which is just a single proton, is the size of a Ping-Pong ball, then its electron would be whizzing around some three hundred meters away. Atoms are large because the electron does not interact strongly. Were the electron strongly interacting, the electrons in an atom would be pulled in much closer to the nucleus by the strong force, and atoms would be much smaller.

THE THIEF OF BAGHDAD

We have seen that whatever passes through a wall had better not interact electromagnetically. Clearly, it had also better not interact strongly lest it feel the strong force exerted by the atomic nuclei in the wall. What we want for the dark matter is a particle that interacts only weakly and gravitationally. Is there such a particle? If so, it would be a true *passe-muraille*, sailing right through the force fields put up by the electrons and nuclei in the wall.

Yes indeed, there is such a particle: the famous and infamous neutrino, postulated in 1933 and discovered in 1955.

To discuss the discovery of the neutrino, I have to explain Einstein's famous formula $E = mc^2$. (I have already mentioned this famous equivalence of energy and mass in Chapter 3, in connection with black holes.) Think of an object's mass as its substance or heft, an intrinsic property of the object. Its energy, on the other hand, depends on how fast it is moving. In Newtonian physics, an object sitting still has no energy. Einstein discovered, however, that even an object sitting still has energy equal to mc^2, the so-called rest energy. (Since the speed of light c is so large compared to the speeds we normally encounter, an enormous amount of energy is locked up in the mass.)

Because of Einstein's discovery, there is a key distinction between massive and massless particles in regard to the minimum amount of energy a particle can have. The minimum amount of energy an electron can have is its mass times $E = mc^2$; in contrast, a massless particle like the photon can have as little energy as you like. For instance, while the primeval photons were tearing nuclei apart, they had energies of millions of electron volts; by now they only have energies of ten thousandth of an electron volt. In the future, they will have even less energy as the universe expands.

Now we are ready to go back to the discovery of the neutrino. By the late 1920s, physicists had understood that radioactivity involves a nucleus in which the arrangement of protons and neutrons is not quite stable. In a given period of time, there is a certain probability that the nucleus will transform itself into a stable nucleus, known as a daughter nucleus. Depending on the nucleus, either the strong, the electromagnetic, or the weak interaction may be responsible for the decay. If the weak interaction is responsible, the probability per unit time for the decay is extremely small—hence the name "weak interaction." As the radioactive nucleus decays into the daughter nucleus, it ejects a particle to make up the energy difference between itself and the daughter nucleus. When the electromagnetic interaction is responsible, a photon is ejected. When the weak force is responsible, an electron is ejected.

Invoking the fundamental law of energy conservation, physicists confidently predicted how much energy the ejected particle should have. The law of energy conservation states that Nature must balance Her energy account: Energy can neither appear from nor disappear into thin air. Thus by simple subtraction, physicists can deduce how much energy the ejected particle should have. Call it E^*.

This expectation was verified in the electromagnetically driven decays: The ejected photon always comes out with the energy E*. But surprise, surprise! It was found in 1927 that the electron ejected in decays driven by the weak interaction does not always come out with the expected energy E^*. In one decay, it might come out slowly; in another, it might come out much faster. Once in a while it would have the energy E^*. (However, its energy is never more than E^*, indicating that E^* is indeed the maximum energy available.)

Where did the missing energy go? The correct resolution of this conundrum was given in 1930 by Wolfgang Pauli. He suggested that a hitherto unknown particle, which interacted neither strongly nor electromagnetically and thus would have escaped detection, carried away the missing energy, like a black-clad thief disappearing into the night.

MASSLESS AND INVISIBLE

The mysterious particle, postulated to interact only weakly (and gravitationally, of course), was later given the Italian name neutrino. When a colleague confused Pauli's hypothetical particle with the neutron, Enrico Fermi had replied that no, it is not the neutron, but sort of a "little neutron."

Pauli also deduced that the neutrino must be massless. Suppose the neutrino has a small mass m. Then the least amount of energy a neutrino sitting can have is equal to mc^2. Since part of the available energy E^* in the decay of a radioactive nucleus must be budgeted to produce the neutrino, an energy of at most $E^* - mc^2$ would be left for the electron. Since the electron does come out with the energy E^* once in a while, Pauli concluded the neutrino must be massless.

Knowing how weak the weak interaction is, Pauli concluded that a neutrino, like a veritable ghost, can pass through the entire earth without interacting. The neutrino is the *passe-muraille* of Aymé. Granted, it is only a particle rather than a macroscopic structure, but it is the best that physicists can come up with.

IN THE DARK

you want to know
whether i believe in ghosts
of course i do not believe in them
if you had known
as many of them as i have
you would not
believe in them either

—DON MARQUIS

The need to believe in ghosts appears to be universal to human societies. Indeed, the ability to conjure up phantoms from the dark and to translate our fear into specific forms may be uniquely human.

I grew up with ghost stories and still sometimes, all alone late at night, I wonder if the darkness is not populated with ghosts. Under the right circumstances, ghost stories can reach deep into our collective human psyche. When I visited Yugoslavia several years ago, a physicist took me for an outing to the countryside. We came back late at night along a deserted country road. My companions took turns telling the local Slavic ghost stories—hair-raising, skin-creeping, nightmare-inducing stories. Believe me, the circumstances were right.

Well, are there ghosts or not? Following Galileo, let me imagine a dialogue between a physicist and a believer in ghosts—say, a modern Hamlet.

H.: Angels and ministers of grace defend us! Be thou a spirit of health or goblin damn'd? Oh, hi there, you are a physicist. Do you believe in ghosts?

P.: As a self-respecting academic, I am not going to start a discussion without first agreeing on a definition. Supposedly, ghosts are capable of doing many things, mostly nasty. But let us focus on their ability to pass through walls. For this discussion, let us just define a ghost as a creature with a human form who can pass through walls. If you can't pass through walls, you are human.

H.: Your definition of a human reminds me of Socrates's definition: a featherless biped. One of Socrates's students was prompted to present the old man with a plucked chicken.

P.: Well, never mind. You know what I meant.

H.: Sure.

P.: I don't know if I believe in ghosts or not, but let me point out that the belief in ghosts entails a serious logical inconsistency. To pass through walls, a ghost has to be made of stuff that interacts neither strongly nor electromagnetically. But if the stuff does not interact electromagnetically, then how can you see the ghost? Ghosts are usually depicted as semitransparent—ethereal, I guess, is the word. But to see it at all, it must interact with the photon, and that means it interacts electromagnetically.

H.: Hmmm . . .

P.: For instance, a neutrino can pass through walls because it interacts only weakly. But surely when people talk about ghosts they have in mind something more than a neutrino.

H.: That's it! You build up a creature using neutrinos. All right, you can't see it, but it is well known that there are ghosts you are not supposed to be able to see, like the kind that specialize in making empty rocking chairs rock.

P.: But there is a catch. To the extent that neutrinos interact weakly with electrons and protons, they also interact weakly with each other. In an atom, the electrons stick around because of their electromagnetic attraction to the protons. In contrast, neutrinos are not attracted to each other. You just cannot build up a structure with neutrinos. And there is another problem. If you want to make a rocking chair rock, then you must exert some influence on the rocking chair. You must interact with the atoms making up the rocking chair.

H.: Perhaps there is an interaction that you physicists do not know about.

P.: It is possible, but any such interactions would have to be extremely weak. Experimenters, using particle accelerators, have carefully searched for hitherto unknown interactions. They all dream of a trip to Stockholm, so they try really hard. But they haven't found any. They set limits on the strengths of various hypothetical interactions. In other words, an experimenter can assert that unless a given interaction is much feebler than such and such, then he or she would have noticed it. Of course, one can consider an interaction involving ordinary particles and a hitherto unknown particle and just say that this new particle is so massive that even the biggest accelerators do not have enough energy to produce it. This kind of hypothetical interaction is harder to rule out. Even so, the indirect effects of this particle may be observable in other physical processes.

H.: Aha, I've got it! You just gave me an idea. What if ghosts are made of new types of particles that interact substantially with each other, so that they would bind with each other into atomlike structures and then into interesting forms of everyday sizes, but that interact very feebly with ordinary particles such as electrons and protons, so that they could have escaped detection?

P.: Yes, that is entirely possible, and your ghosts would be able to pass through walls. I want to make a comment on your scheme, but for the moment let me point out that you are still cornered into saying that ghosts cannot interact much with ordinary particles. These ghosts won't be much fun then. Your ghosts can pass through walls, but because of that very fact, they can't rock rocking chairs and they can't talk to us. You just cannot get around that.

H.: Hmmm . . . But it is still kind of neat that an entire world, with its objects and creatures, can exist without our knowing about it.

P.: Absolutely. Physicists even have a technical term for it: the shadow world. Just imagine doubling every known particle type: With the electron, there is a shadow electron; with the proton, a shadow proton; and so forth. Imagine that the interaction between the shadow electron and the shadow proton is exactly the same as the interaction between the electron and the proton but that the shadow particles interact only feebly, if at all, with ordinary particles. There you have it. You can have chairs and televisions in the shadow world without our noticing it.

H.: That is exactly what I had in mind. By the way, you know that people still are going to think that they can get around the problem that ghosts can't interact if they want to pass through walls. Why can't ghosts switch the electromagnetic interaction on and off?

P.: To some extent, the confusion is caused by the phrase "switching on and off." When you switch off a light, you don't switch off the electromagnetic interaction, of course. You break a contact so that electrons can no longer go around the circuit. It is just like a faucet. To say that the electromagnetic interaction can be "switched" on and off is to say that the fundamental laws of physics can change with time. There is a deep theorem that says that energy would not be conserved if the laws of physics change with time.

H.: You mentioned that you had a comment on my shadow-world scheme. What is it?

P.: Well, as you know from the old man's toy, everything interacts gravitationally—

H.: Wait, I may be in this book, but that does not mean I have read the book!

P.: Well, you should read it. But just let me say that if there is a shadow star or planet floating about nearby, we will certainly feel its gravitational pull.

H.: I suppose that if we were to talk about a particle that does not interact at all with ordinary matter, not even gravitationally, were that possible, philosophers surely would get irritated and start pontificating on what it means to say that such a particle "exists."

P.: Talking about ghosts, I suppose that ghost sightings can most easily be attributed to electrical brainstorms. It is awfully easy to fool the human brain, as you know. Every time we put on glasses, every time we watch television, we are purposely fooling our brains. A particularly interesting example involves those miniature stereos that joggers use. The speakers in these stereos are so small that they cannot accommodate the lowest harmonics of certain instruments, with their long wavelengths. Nevertheless, even accomplished musicians would swear that they hear the note. I can well understand how some people could swear that they have seen ghosts.

H.: People are just going to say that you physicists do not understand the psychic power of the mind!

P.: Sure, anything is possible. We physicists are always careful to limit our studies to the nonbiological physical world so that we know what we are talking about. Eventually physics will have to make contact with human consciousness, but that time has not come yet. There may be psychic power afoot, for all I know, but if it originates in any physical particles, then we can go through our discussion again. For instance, how do these particles, if they are particles yet unknown, stay within the physical confines of our skull?

H.: Good, but I still want to ask you, do you believe in ghosts or not?

P.: It depends on how dark it is.

THE NEUTRINO AS DARK MATTER

You are now probably shouting, "What did I read the first part of this chapter for? What about the neutrino?" See, it is not that hard to solve a cosmological problem. The neutrino is indeed one of the leading candidates for the dark matter particle. It satisfies all three criteria for the dark matter particle.

First, it is massless, so there isn't a particle with lower mass into which the neutrino can decay. Second, neutrinos interact feebly. We could live inside a sea of neutrinos without knowing it. Not participating in the electromagnetic interaction, the neutrino is manifestly dark. Finally, according to the accepted cosmology, the universe is suffused with neutrinos. Let me explain.

COSMIC FOSSILS EVERYWHERE

Shortly after the big bang, neutrinos were produced in abundance. Because of the weak interaction, an electron colliding with a proton can produce a neutron and a neutrino. But you object, "I just learned that the probability of a weak process occurring is very small." Indeed, but the early universe was incredibly hot and crowded with particles. Electrons and protons kept bumping into each other at a tremendous rate. Thus, although the probability that a neutrino is actually produced each time an electron and a proton collide is tiny, neutrinos were produced profusely.

As the universe cooled and expanded, the electrons and protons drifted apart, and the production of neutrinos gradually ceased. Ever since, these primeval neutrinos have been drifting, cosmic fossils floating in the universe. Some of these neutrinos may have encountered lumps of matter here and there and interacted with them. Radioactive decays have added some neutrinos to the universe. But both effects are negligible. The vast majority of the primeval neutrinos will be floating around till the end of time. Physicists have calculated that at the moment there are, averaging over the universe, about a hundred primeval neutrinos per cubic centimeter.

MASS OF THE NEUTRINO SEA

How much mass per cubic centimeter does this sea of neutrinos amount to? Unfortunately, not a hill of beans, as they say. As the universe expands, the energy carried by each neutrino gets reduced to almost nothing. More precisely, the energy of each neutrino has decreased by an enormous factor equal to the expansion factor of the universe since that early epoch when neutrinos were produced in abundance. The reduction of the neutrinos' energies, known as the *neutrino redshift,* is based on exactly the same physics as the photon redshift and the reduction of the primeval photons' energies as the universe expands.

Recall that there is, averaged over the universe, about 1 gram of dark matter in a cube 100,000 kilometers on a side. In that same cube, the present energy density of neutrinos comes to only about 0.0002 gram.

We conclude that even though the neutrino appears to be a ready-made candidate for the dark matter particle, it cannot account for the observed mass density of the dark matter.

The lay reader encountering a scientific statement is often unsure as to where the statement lies in the spectrum between a fact engraved in stone and pure fiction. What about this particularly involved cosmological story about neutrinos being produced in the early universe, drifting through the eons, and now carrying a minuscule amount of energy?

In fact, physicists can tell this story with an unusual amount of confidence because essentially the same story can be told about photons. And the story is no fairy tale because the primeval photons, having drifted through the eons accompanied by the neutrinos like some cosmic hobos with their dogs, had actually been detected about twenty years ago, as we saw in Chapter 6.

THE MASSIVE NEUTRINO

A number of years ago, the physicists R. Cowsik and J. McClelland noticed a loophole in the argument that neutrinos cannot account for the dark matter: The argument assumes the neutrino is massless. Suppose the neutrino has a tiny mass. Once again, recall what Einstein taught us. As a moving particle slows down, its energy decreases. But Einstein differed from Newton by saying that a particle, even at rest, contains an amount of energy equal to $E = mc^2$. This rest energy of a particle is the irreducible minimum amount of energy it can have. Thus, as the universe expands, the energy carried by a neutrino, were the neutrino massive, would keep on decreasing, but it could never get to be less than the rest energy.

But I said all along that the neutrino is massless. How do we really know? As I explained earlier, Pauli deduced that the neutrino is massless because the ejected electron in radioactive decay sometimes comes out with the maximum allowed energy, E^*. But strictly speaking, there can be no absolute statement like this in experimental science. An experimenter can only say that he can verify by an independent means that his device can measure energy to an accuracy of better than 1 percent; therefore, he can assert that the ejected electron sometimes comes out with an energy more than $0.99\, E^*$, and hence the neutrino mass must be less than $0.01\, E^*$.

Cowsik and McClelland made the important point that even if the neutrino has a mass corresponding to a rest energy as tiny as one thousandth the characteristic E^* in some typical radioactive decays, neutrinos can account for the observed dark matter. Since the universe is pervaded with so many neutrinos, the total mass density can be considerable, even if each neutrino carries only a tiny mass.

Ever since the 1930s, experimenters have worked hard to improve the accuracy of their measurements in an effort to detect the mass—if any—of the neutrino. But so far they have been unable either to prove or to disprove the notion that the neutrino may have a small mass.

THE COSMIC WIMP

Meanwhile, particle theorists have been busily inventing various particles to explain the dark matter puzzle. Suffice it to say that these particles are of varying degrees of plausibility, and some physicists are offering bets on them with widely differing odds.

The preceding discussion indicates that the candidate particle must be massive, lest its energy be redshifted to almost nothing. Thus the hypothetical particle comprising the dark matter came to be known by the acronym "wimp," which stands for "weakly interacting massive particle."

Normally, a theorist inventing a particle must first explain why experimenters have never seen the proposed particle. But theorists going into the dark matter business cleared this first hurdle with no sweat, thanks to the very requirement that the wimp interact feebly with known matter.

Physicists regard with suspicion particles invented specifically to solve the dark matter problem. Applying some principle of economy, they generally favor those particles originally invented to solve some other difficulty. For instance, Steve Weinberg and Frank Wilczek, working independently and building upon earlier work of Roberto Peccei and Helen Quinn, invented a particle named the axion for a reason far removed from the dark matter problem. It was realized later that the axion could have the right properties to account for the dark matter. The axion is now regarded as one of the most attractive candidates for the dark matter particle.

DETECTING THE COSMIC PHANTOMS

In the next chapter we will see how detailed astronomical observations may help us determine what the dark matter particle really is.

Meanwhile, a number of experimenters have taken up the challenge of detecting these cosmic phantoms here on earth. Wait a minute, you say, the whole point of the dark matter particle is that it barely interacts. How can it be detected? True, but that's partly compensated for by the sheer number of dark matter particles: We are swimming in a sea of them. Once a detector is built, dark matter particles will be streaming through it day and night without cease. We only have to hope that a few will interact with the detector, leaving some telltale tracks. Thus, with ever more refined techniques and sophisticated technology, the terrestrial detection of the dark matter particle is not inconceivable. One day we may catch one of these ghostly phantoms.

11

Crowned with a Halo

DARK MATTER AND GALAXY FORMATION

In a mystery, until the culprit has actually been apprehended, the detective can only examine the scene of the crime for clues. Similarly, until the experimenters actually catch the wimp, physicists can only study the wimp's effect on the universe. Since the dark matter outweighs ordinary matter, the physicist detectives reason, the wimp must have had enormous impact on galaxy formation.

As we saw in Chapter 8, the early universe was uniformly filled with matter. Inevitably, there were fluctuations or irregularities in the distribution of matter: Here and there were regions with a somewhat higher density of matter than the surrounding regions. The matter in these regions, being denser, would pull matter in from the surrounding regions by the force of gravity. As a result, these regions became even denser with matter, while the surrounding regions became even sparser. The process went on, leading eventually to the formation of galaxies, stars, and other structures out of the primeval haze.

But galaxy formation did not occur until after the formation of atoms. Before that, photons acted as cosmic Robin Hoods, rushing about smoothing out the distribution of matter, as we saw in Chapter 8.

THE RICH BEGET THE RICH

How does dark matter modify this story, told by textbooks published as recently as several years ago? The dark matter allows the rich to get around the photon Robin Hoods!

175

Since the wimps do not interact with photons, unlike the charged protons and electrons, they could start gathering gravitationally long before atom formation. Regions that happened to have a somewhat denser distribution of wimps could start getting even denser by pulling in wimps from the surrounding regions.

Meanwhile, the photons ignored the wimps as they struggled mightily to smooth out the distribution of ordinary matter. All that labor would prove to be in vain. The wimps were already condensing into lumps of dark matter, which tugged at the ambient ordinary matter, urging the protons and electrons to fall in. As soon as atoms formed, ordinary matter fell into the ready-made dark matter lumps. Not only do the rich get richer, but the rich also can count on inheriting from the wimps.

You may be confused by all this talk of dark matter lumps. Earlier I said that the dark matter was more or less evenly distributed. It is a matter of degree: The dark matter is much more evenly distributed than ordinary matter. The universe is suffused by an essentially uniform distribution of dark matter. But here and there are regions with a higher density of dark matter or lumps of dark matter. Within these lumps lies ordinary matter, concentrated into tiny balls.

THE ORIGIN OF IRREGULARITIES

I should emphasize that in this story we still have to assume an initial fluctuation or irregularity in the distribution of the dark matter. The discussion here does not address the logically separate question of where the initial fluctuations might have come from. As we saw in Chapter 9, cosmic strings pulling on the dark matter may provide an answer to that question. The inflationary scenario may provide another possibility. To explain this, I have to digress a bit.

With the passage from classical to quantum physics in the 1920s, physicists had to replace classical determinism with quantum uncertainty. Looking at an electron in an atom, they could no longer say that the electron is here and not there. Instead, they could only state that the probability is 60 percent, say, that the electron is here, and 40 percent that it is not here. Physicists have been reduced to bookies, posting odds on the various possibilities. The stately waltz of classical physics was replaced by the jitterbug of the quantum.

For our purposes here, we only have to know that the deterministic description of classical physics was replaced by a probabilistic description. But a probabilistic description implies fluctuations. With classical physics, we are free to imagine a perfectly smooth distribution of matter. In contrast, with quantum physics, we can give only the probability that the density of matter here is the same as the density of matter there. Quantum physics necessarily requires fluctuations in the distribution of matter in the early universe.

But wait, you say, isn't it true that quantum physics is needed only for describing the world on microscopic scales of atomic dimensions or less? Yes, that's true. On larger scales, classical physics is adequate as the quantum probabilistic description is smoothed over to give an apparently deterministic description.

Thus, cosmologists had long thought that they could simply forget about quantum physics. The fluctuations in the distribution of matter required by quantum physics occur on the subatomic scale. Viewed on the enormous scales of cosmology, these fluctuations are totally negligible.

This was the situation until the inflationary theorist came along. He stretched tiny regions of the early universe out into enormous regions and thus the quantum fluctuations on microscopic scales are also into fluctuations on macroscopic scales. *Voilà!*—the primeval fluctuations and irregularities needed to precipitate structures out of the void.

Just think, quantum fluctuations inflated could have been responsible for the formation of dark matter lumps, of galaxies, and ultimately of you and me! This fascinating idea appeals to our sense of economy, since we are already stuck with quantum fluctuations. There is no need to invent some other source of fluctuations.

Unfortunately, a calculation showed that the fluctuations come out to be much too large. Physicists who believe in the inflationary universe are hard at work modifying the basic scenario so that the fluctuations would come out right.

PLENTY OF TIME TO GET RICH

In what way does the story of the universe filled with dark matter improve over the old one? The new story gets around a problem mentioned in Chapter 8 and that the rich often feel acutely: It takes time to become even richer.

Recall from Chapter 8 that recent observations on the primeval photons coming from different regions of the universe indicate that the distribution of photons was exceedingly smooth when atoms formed. This in turn indicates that the fluctuation was too small for galaxies to have grown in the time available without the early start provided by the dark matter.

Thus these observations provide indirect support for dark matter: With dark matter present, the formation of structures in the universe could start earlier, without waiting for the formation of atoms.

Notice that the existence of the "not enough time" problem also constitutes evidence against the supposition that the dark matter consists of ordinary matter hidden somehow in gas, dust, rocks, or planets. For the dark matter to start condensing before atom formation, it must not interact with photons at all.

In the new story, the universe before atoms formed contained both roughly spherical lumps of dark matter here and there and ordinary matter distributed uniformly. After the ordinary matter forms into atoms, these atoms start to fall into the nearest lump of dark matter. As they fall in, they crash into each other, emitting photons that carry away the atoms' energies. Ordinary matter dissipates and collapses in much the same way as in the old story.

In some of these lumps, matter is squeezed so tightly together that some atoms are stripped of their electrons. Once again, the protons and neutrons are able to get close to each other, just as in the good old days when the universe was hot. They get so close together that the strong force takes over, igniting nuclear processes. Suddenly the universe sparkles with zillions of star fires. Nuclear reactions proceed, for the second time in the history of the universe. Were these protostars able to predict the future, they might foresee that the energy and warmth provided by this second nuclear ignition will allow civilizations to evolve and learn how to ignite nuclear reactions on their own.

CROWNED WITH A HALO

The behavior of the wimps differs crucially from that of ordinary matter. To a good approximation, particularly when compared with the atoms of ordinary matter, the wimps do not interact, either with each other or with atoms. Thus they do not collide, and they do not lose energy. They do not participate in the dissipative spree of ordinary matter. They continue to drift lazily along, on orbits determined by each particle's energy and the gravitational force acting on it. (In the same way, to the extent that frictional forces on the earth are small, the earth is not going to crash into the sun.) The particles of dark matter are like a swarm of gnats buzzing around the collapsed ordinary matter. Each of the visible galaxies has a puffed-up ball of dark matter around it. Technically, the puffed-up ball of dark matter is known as a halo.

Considering that our daily experience is full of friction and collapse, not to mention pain, we may not appreciate fully this crucial difference between the behavior of dark and ordinary matter upon first hearing. We see energy dissipated all too readily and thus tend to forget that a way of dissipating energy may not always be available. When a physical structure collapses in everyday life, energy is carried away by sound waves, by electromagnetic waves in some extreme cases, and by the ambient air eventually soaking up the heat generated. In the empty silence of space, however, particles that do not interact electromagnetically can only keep on waltzing around each other. Even though they influence each other's motion via gravity, they cannot come together.

TWO DIFFERENT COSMIC SCENARIOS

I would like to give you a taste of current research in dark matter. A clearly important issue is the size of the dark matter lumps that formed early on. As we will see presently, the size depends on how fast the wimps were moving in the early universe, which in turn depends on the nature of the wimp.

If we are told that space contains a certain amount of dark matter, we can think of two extreme possibilities. The wimp may have a tiny mass, and each volume of space contains lots and lots of this particle. This possibility is exemplified by the neutrino if it has a mass. Or the wimp may be very massive and each volume of space contain correspondingly fewer wimps. To sharpen their thinking, physicists find it useful to focus on these two extreme possibilities.

RUSSIAN PANCAKES

Suppose that the neutrino has a small mass and that the dark matter consists of neutrinos. In the early universe, neutrinos had lots of energy. Since the neutrinos mass is tiny, each neutrino had energy far in excess of its mass. Thus the neutrinos were zinging around almost at the speed of light. This has the important consequence that small lumps of neutrinos quickly got smoothed out. The neutrinos in the lump were rushing off in all directions, barely interacting with each other. Thus a given neutrino quickly found itself outside the lump, perhaps on its way to join another lump.

In contrast, it would take a while for a neutrino somewhere in the center of a really huge lump to make its way out. The larger the lump, the longer it lasts, obviously.

As the universe cooled, the energy of each neutrino was reduced steadily according to the redshift phenomenon mentioned earlier. At some point the typical energy of the neutrinos became comparable to the neutrino mass. Suddenly the neutrinos no longer moved at close to the speed of light. They began to dawdle and stick around. Clearly, we want to know the size of the smallest lump that could have stayed together just before the neutrinos slowed down. Lumps smaller than this characteristic size would not have survived. Lumps bigger than this size, in contrast, survived and went on to initiate galaxy formation, as described earlier. (Figure 11.1.)

It should be clear that since the neutrinos were zinging around almost at the speed of light, this characteristic size would turn out to be enormous. A rather simple calculation, taking into account the expansion of the universe, shows that this characteristic lump is indeed enormous

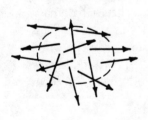 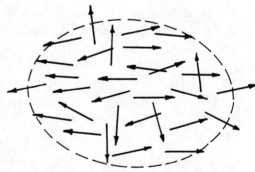

11.1. *The length of the arrows represents the average distance a neutrino travels in a characteristic period of time. The smaller lumps disperse rapidly, while the big lumps stay together.*

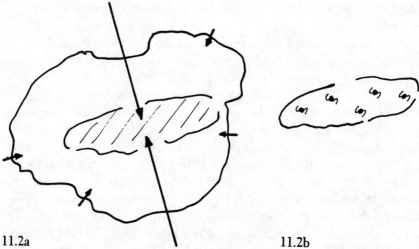

11.2a   11.2b

11.2a. When an irregular lump of matter collapses, the rate of collapse would in general be different in different directions. The longer arrow indicates the direction of fastest collapse. The end result is a pancake-like object. **11.2b.** *The ordinary matter in the pancake fragments later into galaxies.*

and contains enough mass to make about a million galaxies, assuming that the neutrino has a mass as big as that currently allowed by experiments.

This scenario, known as the fragmentation or pancake scenario, has been studied particularly by the eminent Soviet physicist Y. B. Zel'-dovich and his collaborators. As the ordinary matter contained in one of these giant lumps of dark matter collapses, the collapse rates along the three perpendicular directions of space are bound to be different. (See Figure 11.2.) Thus the ordinary matter tends to collapse into a more or less flat object, looking something like a cosmic pancake. The pancake would then fragment later into a large number of galaxies by a rather poorly understood process.

THE AGGREGATION SCENARIO

In contrast, if the wimps were heavy, then at the relevant epoch they were already moving slowly. Even rather small lumps would stay together. The wimps contained in these lumps were moving too slowly to escape gravity's clutch.

As time passed, these small lumps, because of their gravitational attraction for each other, merged into bigger lumps, which in turn merged into even bigger lumps. Thus were galaxies born.

This scenario, known as the aggregation or hierarchical scenario, has been championed by the eminent Canadian-American cosmologist James Peebles, among others. In this scenario, the growth of structures proceeded from the small to the large. After their birth, galaxies attracted each other gravitationally, and thus clusters of galaxies formed naturally. These galactic clusters, in turn, formed into superclusters. The universe is said to be built from the bottom up. In contrast, in the pancake scenario, huge superclusters formed first and then fragmented into clusters and galaxies. In that scenario, the universe is said to be built from the top down.

TOP DOWN VERSUS BOTTOM UP

Controversies are raging between subscribers to one or the other of these two possible scenarios for the universe. (Naturally, there are also some fence-straddlers who believe in something in between.)

Astronomically, one should be able to distinguish between the pancake and the aggregation scenarios. In the aggregation scenario, galaxies should be older than galactic clusters and superclusters. In the pancake scenario, the opposite should be true. Thus the debate centers on the relative ages of galaxies versus clusters and superclusters.

To answer these questions, one has to study the distribution of galaxies over large scales. It is a mind-boggling field in which galaxies, each with their 10^{11} or 10^{12} stars, are treated as mathematical points. To observe superclusters in the process of formation is to see deep into space. The observational efforts involved can rightly be called heroic.

From these observations, circumstantial evidence has emerged favoring *both* the top-down and the bottom-up universe. We will mention only two pieces of evidence.

The exciting discovery of *cosmic voids* a few years ago appears to support the "top-down" universe. These voids are enormous regions in space apparently empty of galaxies, or at least of bright galaxies. The largest such void known, the great void in Boötes, measures an astonishing 200 million light-years in size. Such voids may emerge quite naturally in the pancake scenario. If galaxies formed in roughly two-dimensional pancakes, the space between the pancakes would be expected to be more or less empty of galaxies. (See Figure 11.3.) In contrast, in the aggregation scenario, one might expect the galaxies, aside from their tendency to cluster, to be distributed more or less uniformly.

But other evidence favors the bottom-up universe. When we look around the universe, superclusters appear to be just forming, as indicated by studies of the motion of galaxies and galactic clusters. In our "immediate" neighborhood, the Local Group, the cluster to which our own Milky Way belongs, is observed to be moving away at about a thousand kilometers per second from the neighboring Virgo cluster, after the motion due to the expansion of the universe has been subtracted. This observation suggests that the Local Supercluster, to which the Local Group and the Virgo Cluster belong, has barely formed, if at all. Some of its constituents still are moving apart.

The evidence is fragmentary and difficult to interpret. Recently, Peebles weighed all the evidence and pronounced in favor of the bottom-up universe. But others disagree.

From the observational evidence, physics have tried to decide what the dark matter particle can be. Some say the neutrino is ruled out. Others say no. The subject is far from settled.

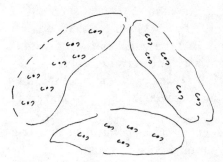

11.3. *The universe is full of holes: If galaxies tend to condense within large pancake-like structures, then the space between the pancakes would be more or less devoid of galaxies. As long as we are on an alimentary tack, we might say that the pancake scenario produces a Swiss cheese universe.*

HOW MUCH MATTER IN THE UNIVERSE?

The amount of dark matter in the universe presents another profound mystery. As we have seen, averaged over the universe there is about 1 gram of dark matter per 10^{30} cubic centimeters. Is that a lot?

To make sense of this question, we must first get away from grams and centimeters, two arbitrary human measures made up by committees of French revolutionaries. We must express the observed density of dark matter in some way intrinsic to the universe. Happily, Einstein's theory provides just such an intrinsic measure. If the universe contains a small amount of matter, it will continue to expand forever—the universe is said to be open. Space is curved outward like the surface of an infinite saddle, as I mentioned in Chapter 2. But if the universe contains too much matter, then the gravitational attraction of the particles of matter for each other will eventually arrest the expansion of the universe. The universe will expand up to a certain maximum extent and then proceed to contract. The universe is said to be closed. Space is curved inward like the surface of a sphere. One can therefore define a critical mass density such that if the universe contains less mass per unit volume than this critical mass density it would be open, and if it contains more, closed.

The critical mass density provides a natural unit of measurement for the mass density of the universe. In other words, instead of saying what the density of the universe is in some dumb units like grams per cubic centimeter, cosmologists simply compare the density of the universe to the critical mass density. They denote the ratio of the actual density of the universe to the critical mass density by the Greek letter Ω (omega). Thus, if Ω is less than 1, that indicates that the actual mass density is not quite large enough to curl up the universe. The universe is open. If Ω is greater than 1, the universe is closed. If Ω happens to be exactly 1, then the universe is just on the borderline between being open and being closed.

Before I get back to the amount of dark matter in the universe, I should mention that for a long time physicists have argued over how much matter the universe should contain. A priori, Ω could have been any number. It could have been 0.00001, say, or just as reasonably, 10. But some physicists felt deep down that somehow the value $\Omega=1$ must be special, so they said that Ω must actually be 1. Since God does not have any particular reason to prefer an open universe over a closed universe,

184

it would make sense, some physicists argue, for Him to choose the universe to be precisely in between. (Figure 11.4.)

The notion that Ω should be equal to 1 remained a mere hunch until the theory of the inflationary universe came along. It turns out that that theory also predicts $\Omega = 1$.

A crude way of understanding this is to picture inflation stretching space out. Take a curved surface made of extremely elastic rubber and stretch it out by some huge factor. You have effectively made it flat. Similarly, after inflation, space is curved neither inward nor outward. There is a certain irony that after all the discussion of curved space in Einstein's theory, the universe on a large scale may turn out to be flat.

Having gone through this long background, let us get back to the question of whether there is a lot of dark matter in the universe.

The observed amount of dark matter corresponds approximately to $\Omega = 0.2$. In other words, the amount of dark matter amounts to one fifth of what is needed to close the universe.

11.4. *Physicists arguing about how much mass is in the universe.*

This fact has generated a lively debate. Everyone agrees that it is remarkable that Ω turns out to be so close to 1. Those who believed all along that Ω must be 1—let us call them believers—are gratified. But the nonbelievers, those who did not believe that Ω must be 1, are equally pleased. "Ω is not 1, but 0.2," they point out. "No, no, Ω is really 1," the believers insist. "The measurements showing Ω is 0.2 are just off." "Nonsense," scoff the nonbelievers. "Show us how the measurements could have gone wrong."

The believers slouch off, but soon they come back pointing out that the conclusion $\Omega = 0.2$ depends on an implicit assumption. When we look deep into the night sky, we see points of light. Until the advent of dark matter, there was no reason at all to question the assumption that the points of light correspond to concentrations of mass. This assumption, that light traces mass, is in some sense basic to astronomy. But now, with dark matter lurking about and dominating ordinary matter, we can no longer be so sure. We only know that ordinary matter is concentrated about the points of light. There could be regions dense with dark matter but somehow with hardly any ordinary luminous matter. If so, the actual mass density could be higher than $\Omega = 0.2$. Perhaps it could even be consistent with the cherished folk belief that $\Omega = 1$. Put that in your pipe and smoke it, you nonbelievers!

In principle, the distribution of dark matter may be determined directly if gravity-wave astronomy, in contrast to electromagnetic-wave astronomy and as described in Chapter 3, is ever realized. A big lump of dark matter, although it does not emit any electromagnetic wave by definition, may be emitting gravity waves as it sloshes about.

The issue of how much mass the universe contains is hotly debated. Think about it—it is incredible. These creatures, who emerged just recently on this speck of dust, are now arguing about the mass of the universe to within a factor of five!

COSMIC HUMILITY

The dark matter problem has expanded our horizons and generated a thriving area of research. Physicists from around the world gather to argue for their favorite wimp. Experimental evidences for the neutrino mass are scrutinized. Astronomical observations are mulled over. Is the universe top down or bottom up? The atmosphere is unusually ecumenical: People studying the extremely small talk to people studying the extremely large. The mystery probably will linger on for years, but we hope we will find out eventually who the ghost riders in the sky are.

Here is a summary of the clues: (1) Motion of astronomical objects indicates almost incontrovertibly that the universe is pervaded with an invisible stuff, called the dark matter. (2) The dark matter outweighs visible ordinary matter by roughly a factor of ten. (3) Various arguments show that the dark matter cannot be ordinary. (4) If the neutrino has a mass, it may be the wimp making up the dark matter. (5) The dark matter drastically revises our theory of galaxy formation, allowing structures to condense earlier than previously thought possible. (6) Depending on the nature of the wimp, the universe may be top down or bottom up. (7) The dark matter mass density is close to the critical mass density needed to close the universe.

The long history of our growing understanding of the physical world has been described sometimes as a humbling process, a steady erosion of anthropocentrism and geocentrism. The ancient Chinese thought their Middle Kingdom occupied the center of the world. Perspectives had to be broadened. The Greek Anaxagoras was ridiculed for suggesting that the sun may be as large as the Peloponnesus. Anthropocentrism and geocentrism eventually gave way to heliocentrism. The belief that the sun is at the center of the galaxy persisted into this century. It was only in 1915 that Harlow Shapley determined that we are out near the edge. For years afterward, astronomers believed that ours is the only galaxy, thinking that what we now recognize as other galaxies were merely clouds of luminous gas within our galaxy. We now know better: There appear to be as many galaxies in the visible universe as there are stars in our galaxy.

But just as we finally recognize ourselves as passengers on a smallish planet circling an insignificant star lost somewhere near the edge of an ordinary-looking galaxy drifting inside a relatively sparse cluster of galaxies in some region of the universe resembling any other region, we learn that the matter out of which you and I and stars and galaxies are made may not even be the main component of the universe. How humble do we have to be?

THE MYSTERY OF GRAVITY

We ponder the mystery of falling and our present struggle to understand gravity, thus entering into terra incognita.

The Fall and Rise of Gravity

THE TIPSY BUS

When I was a child, I came across a book in which there was a picture of a double-decker bus tilting at a precarious angle. The author explained that physicists can actually calculate how far the bus can be tilted before tipping over. This passage impressed me deeply, partly because I was taken by the notion that we can predict how Nature will behave, and partly because I had ridden some weeks earlier on a double-decker bus—no, not only that, but I had sat in the front row on the top deck—and the thought that it might tip over was deliciously titillating.

To this day I remember quite clearly that it was late afternoon and the light of day was just starting to fade. I set the book down and wandered into my parents' bedroom, where I found on my mother's dressing table a plastic box with roughly the proportions of a double-decker bus. I experimented to see how far I could tilt the box before it would tip over. According to the book, the bus would tip over if the vertical line that passes through the center of the bus passes outside the base of the bus. (See Figure 12.1.) At that age I was a skeptic, and I was eager to prove the book wrong. But soon I was convinced and fascinated: Nature is describable by simple rules!

THE FALL OF GRAVITY

As is the case with most physicists, my first exposure to physics as such was to the mystery of gravity. Indeed, as soon as we are born,

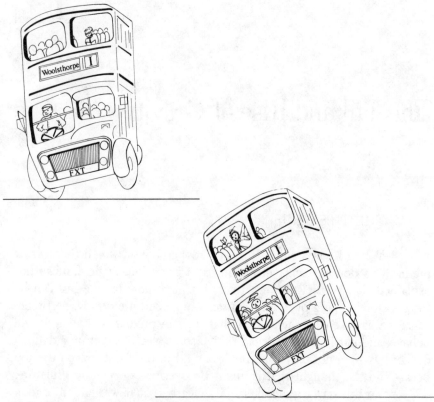

12.1. *Whether the bus would tip over depends on whether a vertical line drawn from its center of mass passes through its base. Aside from showing that gravity acts vertically, this Figure does not illustrate the nature of gravity as much as it illustrates the mechanics of rigid objects.*

before we even open our eyes, we become aware of gravity. We lose our handholds, and we drop out of trees. We trip and we fall. What do these and myriad other everyday experiences tell us about the essence of gravity?

Newton recognized gravity to be a force, and he quantified it. But it took the old man's toy for us to glimpse the profound mystery of gravity with its astonishing connection to space-time itself. From the old man's toy to the expanding universe, we have seen gravity at play and at work. In this chapter I will tell you how contemporary physicists have been relentlessly trying to get at the heart of gravity.

Physics began with gravity but may also end with gravity. Of the four forces, gravity is by far the least understood. The reason is gravity's incredible weakness, as was discussed in Chapter 3.

Is Einstein's theory the final word? Physicists have dared to consider various modifications of Einstein's theory. Do you think that the deviation between Einstein's theory and Newton's theory is minuscule? Compared to these pitifully small effects, the deviations between Einstein's theory and its variants are smaller yet by another huge margin.

The days of dropping lead weights off the Leaning Tower of Pisa are long gone. Progress in our understanding of gravity had been hampered by the virtual impossibility of doing experiments. With the lack of progress, interest in gravity slackened.

The weakness of gravity also implies that it does not play much of a role unless astronomical masses are involved. In their studies of physical phenomena, from the huffing and puffing of the steam engine to the dance of electrons in microchips, from the propagation of electromagnetic radiation across the universe to the annihilation of quarks by antiquarks, physicists can pay not the slightest attention to gravity. Even in studying the life and death of stars, physicists tend to think of the stars' nuclear metabolism, barely keeping in mind that gravity is needed to hold the stars together. And in our tour of the universe, while we know that gravity is essential, we do not often see it on center stage.

RELEGATED TO A BIT PART

And so gravity, a star in the first act of the drama of physics, was soon relegated to a bit part, or at best called to a supporting role at times. To be sure, gravity did have its moments in the limelight. Early in this century, Einstein's theory of curved space-time and the discovery of the expanding universe brought gravity once again to center stage, but only for a while. With the coming of the quantum, the subatomic and subnuclear world captured physicists' attention. Our graduate curriculum reflects this insignificance of gravity in the collective consciousness of the physics community. Typically, graduate students are required to take two or even three years of quantum physics, while Einstein's theory of gravity is brushed over in a half-year optional course. Plenty of physics Ph.D.'s never bothered to learn about gravity. That was certainly the attitude when I started out in physics. Einstein's child was admired but neglected.

But gravity is not to be denied. If it doesn't get more stage time, it's gonna call its agent! The audience hears ominous rumblings from the wings. In its petulance, gravity is threatening to disrupt the entire play.

A MARRIAGE PROPOSAL SPURNED

The rumblings have gone on for a long time. Just as Einstein steadfastly refused to subscribe to quantum physics, so his theory has steadfastly resisted marriage with the quantum. No, no, I don't want to marry a quantum. I wanna be a star!

Quantum physics is not so much a theory as a principle that when combined with an existing classical theory turns it into the corresponding quantum theory. Combine classical Newtonian mechanics with the quantum principle, and we obtain quantum mechanics; combine classical electrodynamics with the quantum principle and we obtain quantum electrodynamics, and so on. In each case, physicists must check that the resulting theory makes sense. If it does make sense, they say that the theory is renormalizable. Crudely speaking, this terminology indicates that the theory can be made "normal" once again after its passage from the classical to the quantum. For instance, quantum electrodynamics was proved to be renormalizable in the early 1950s.

But when classical gravity, as described by Einstein's theory of gravity, is combined with the quantum principle, the resulting theory of quantum gravity does not make any sense. Quantum gravity is nonrenormalizable. For a while, the weak force, the force responsible for certain radioactive decays of atomic nuclei, also appeared to be nonrenormalizable. At least in this respect, gravity had company. But then, around 1971, physicists discovered that the weak force is renormalizable after all. Of the four fundamental forces known to physics, the strong, the electromagnetic, and the weak are all now known to be normal. Gravity alone remains abnormal. Agh, for that I really will refuse to play!

WHO IS SPURNING WHOM?

Some physicists simply shrug and say that perhaps the quantum principle is incomplete. I for one, if forced to bet, would be inclined to think that the quantum principle is but an approximation of a deeper truth. In the failed marriage between the quantum and the theory of gravity, who is spurning whom?

Another possibility is that gravity as we know it is not a fundamental force. Certainly, nobody would insist on applying the quantum principle to the equations describing the flow of liquids. If one does insist, the resulting theory will be nonrenormalizable. But who cares? We know that the forces in fluids are not fundamental but merely manifestations of the underlying electromagnetic forces between the fluid of atoms. Perhaps, in much the same way, gravity is merely the manifestation of a deeper underlying physics.

I mentioned in Chapter 3 that space-time, with its warping and unwarping, is an elastic medium just as a piece of Jell-O is an elastic medium. Just as the forces governing the flow of fluids are not fundamental, so, too, the elasticity of Jell-O merely reflects a deeper reality. Various physicists, starting with Andrei Sakharov and including Steve Adler and me, have developed this thought, ascribing the elasticity and curvature of space-time and hence gravity to some underlying dynamics.

But these are decidedly unpopular attitudes. There is no evidence the quantum principle is incomplete or that gravity is not fundamental.

REFUSAL TO JOIN IN

As time went on, the petulant refusal of gravity to join in the play became ever more embarrassing. In the 1970s, the other three fundamental forces were unified into one single force, as was mentioned in Chapter 7. Physicists discovered that the strong, the electromagnetic, and the weak forces, despite their apparent differences, are in fact related on a deeper level. The three forces represent different manifestations of one single underlying force. Gravity is excluded. (Figure 12.2)

In his definitive biography of Einstein, Abraham Pais wrote that he was struck by the apartness of the man. The gravitational interaction, in so many ways Einstein's child, also stands apart.

In a deep sense, gravity is really different from other three forces because of its intimate connections with space-time. Physicists always have thought of space-time as the stage on which particles and the forces between them play. But gravity is curved space-time, and curved space-time is gravity. We spoke figuratively of gravity grumbling in the

12.2. *Gravity refuses to join in the dance.*

wings, waiting for its chance on center stage. In fact, gravity is the stage!

Physicists are excited and challenged by the mystery of gravity because a deep understanding of gravity would surely reveal the very heart of space and time.

STRONGER AT HIGHER ENERGIES

Thus gravity, too feeble to be of much weight and branded as abnormal for its refusal to marry the quantum, stood apart from the other three forces. But it plots its return, like a fading movie star of yesteryear. The day will come when it will come roaring back!

We are getting carried away in our use of metaphors. The phrase "the day will come" suggests a temporal progression. Actually, the progression is in energy. The existing theory of gravity indicates that gravity will not be feeble forever. As we study physics at ever-increasing energy scales, gravity will become ever stronger.

That gravity will become stronger at higher energies is so remarkable that it is well worth understanding.

According to Newton and as discussed in the Prologue, the gravitational force between two particles of masses m_1 and m_2, respectively, is equal to the product of the two masses multiplied by Newton's constant G and divided by the square of the distance d between the two particles:

gravitational force $= (G\ m_1\ m_2)/d^2$

For comparison, the electric force between two particles with charges e_1 and e_2, respectively, is equal to the product of the two charges divided by the square of the distance d between the two particles, that is,

electric force $= (e_1\ e_2)/d^2$

Specifically, let both particles be protons. Then the quantity $(G\ m_1\ m_2)$ is some 10^{36} times smaller than the quantity $(e_1\ e_2)$. The number 10^{36} expresses the extreme weakness of the gravitational force compared to the electric force.

In Newtonian physics, this would be the end of the story. One quantity in Nature is measured to be fantastically smaller than some other quantity, and that's that. But according to Einstein, mass is equivalent to an amount of energy equal to mass times the speed of light squared: mc^2.

The equivalence of mass to energy suggests that the gravitational force between two objects ought to be proportional to the product of their energies and not to the product of their masses. Newton's formula should be modified. In other words, in the formula $(G\ m_1\ m_2)/d^2$ for the gravitational force between two objects, we should replace m_1 and m_2, the masses of the objects, by E_1/c^2 and E_2/c^2, the energies of the objects divided by the speed of light squared. Thus the gravitational force between two objects is in fact proportional to $(G\ E_1\ E_2)$ and not to $(G\ m_1\ m_2)$, as Newton thought.

Ordinary objects and even astronomical objects move so slowly compared to the speed of light that the energies associated with their motions are completely negligible compared to the energies contained in their masses. Thus the replacement of masses by energies in the formula for the gravitational force makes only a tiny difference. Theoretical physicists, however, are free to contemplate the gravitational force between two arbitrarily energetic particles. The replacement of masses by energies can then make all the difference in the world.

Consider two energetic protons. As the energies of the two protons increase, so does the gravitational force between them. Gravity will not be weak forever!

To keep things simple, let us suppose that the two protons have roughly the same energy. First, just a reminder that the physicists in this business like to measure energy in GeV (which is the energy an electron would have if accelerated through a voltage of a billion volts). If you plug the proton's mass into $E = mc^2$, you would find that the energy contained in the proton's mass amounts to approximately 1 GeV (which explains, of course, why it is convenient to use GeV as the unit of measurement rather than calories, say). Thus a proton with an energy of 10 GeV has approximately ten times the energy of a proton sitting at rest. Nothing profound here. I just have to mention all this so we will know exactly what we are talking about.

Back to the two protons. Having replaced masses by energies in Newton's formula, we now have the gravitational force between two protons increasing as the product of their energies. Suppose each proton has an energy of 10 GeV. Then the strength of the gravitational force would have increased by a factor of 10 times 10—that is, a factor of 10^{1+1} = 10^2, or in words, a factor of one hundred compared to what it was when the two protons were sitting quietly. Suppose each proton has an energy of 100 GeV = 10^2 GeV. Then the gravitational force would be 100 × 100 = $10^2 \times 10^2 = 10^{2+2} = 10^4$ times stronger than when the protons were sitting. And so it goes. The gravitational force between the protons gets stronger and stronger as their energies increase. Meanwhile, since the electric charge of a particle is fixed once and for all, the electric force between the two protons does not change as the protons become more energetic. (Actually, I must say in fine print that the electric force does change, but by a factor utterly negligible in this context. The essential point is that the gravitational force, unlike the electric force, is generated by mass and hence energy.)

Now suppose each proton has an energy of 10^{18} GeV. The

gravitational force would have increased by a factor of 10^{18} times 10^{18}—that is to say, a factor of $10^{18+18} = 10^{36}$ compared to when the two protons were sitting quietly. Recall that the electric force between two protons just sitting there and chatting is about 10^{36} times the gravitational force between them. Thus we conclude that when the two protons have energies equal to 10^{18} GeV, the gravitational force between them becomes comparable in strength to the electric force. Clearly, that such an enormous energy is needed is just the flip side of the statement that the gravitational force is 10^{36} times weaker than the electromagnetic force.

ONCE AN ARNOLD SCHWARZENEGGER

Physicists can only imagine an energy of 10^{18} GeV. To appreciate the enormity of this energy, note that at the world's largest accelerators protons have been accelerated up to an energy of only several hundred GeV.

As we go farther and farther back in time in the early universe however, we will eventually reach any energy, no matter how high. Thus, we who live in this now-frigid universe can venture to imagine the universe when its temperature was an almost inconceivable 10^{18} GeV. The universe when Gamow cooked the nuclei of matter, even the universe when primeval energy begat matter, felt tepid by comparison.

Somehow I think of the movie poster of *One Million Years B.C.* that I admired so much as a kid. You know the one, with a scantily clad Raquel Welch and prehistoric beasts charging across the earth. In this early, early universe, 10 zillion zillion years B.C., gravity charges across the universe, knocking heads together like we have never seen it do ever since. Gravity was the real prehistoric monster.

Indeed, you may say, why not imagine the universe at even higher temperatures? You are right—there was even an epoch when gravity was king, the most powerful force in the universe, when monster gravity ruled the universe, as a movie poster might say. It is as if those body-building ads one sees at the backs of comic books have gone berserk: The hundred-pound weakling was once like Arnold Schwarzenegger! As the universe evolved, his muscles turned flabby, and he lost all his strength. (Figure 12.3.)

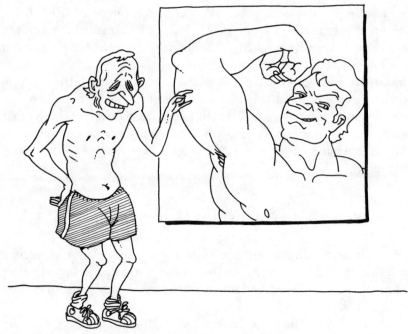

12.3. *As the universe evolves, a strong man turns flabby.*

QUANTUM FLUCTUATIONS

The discussion is incomplete, and strictly speaking, incorrect, without introducing the quantum. Remember the quantum, the character whom gravity refused to marry? We are now in a position to understand why she refused.

Consider again two protons sitting quietly at rest. Classical physics tells us that the electric force between them is given by some value. To the extent that classical physics is but an approximation to quantum physics, that value is not exactly right.

Since I can hardly give you a course on quantum field theory here, I can only try to convey to you the spirit of the argument showing how quantum physics modifies the force between two particles.

In the preceding chapter I mentioned how the probabilistic description of quantum physics necessarily leads to fluctuations. According to Heisenberg's uncertainty principle, we can never be sure that the protons are really sitting still. We instruct two children to sit quietly. As soon as we leave the room, they are at each other's throats and tearing up the room, but the moment we come back, they are again sitting like two little angels. This analogy does not do justice to the full subtlety of quantum physics, but it does suggest, at least vaguely, what goes on in the microscopic world. (As the father of two boys, I know something else wrong with the analogy: When I come back into the room, they will still be at each other's throats.) As soon as we are not observing the two protons, they, too, are shaking every which way, driven by the mad dance of the quantum. Physicists describe the microscopic world as driven by quantum fluctuations. (Strictly speaking, the electromagnetic field between the two protons is also fluctuating. Figures 12.4a and 4b.)

Physicists can calculate the effect of quantum fluctuations using their understanding of quantum physics. They find that the electric force between two protons is equal to the classical value plus a tiny quantum correction. (The quantum correction had better be tiny, for otherwise we wouldn't have believed in classical physics for so long.)

By measuring the electric force between the electron and the proton in a hydrogen atom to great accuracy, physicists have verified that the electric force is indeed modified slightly. (Clearly, our discussion applies to the electric force between any two charged particles.) The detailed agreement between theory and experiment contributes to our confidence in the reality of quantum fluctuation.

12.4a

12.4b

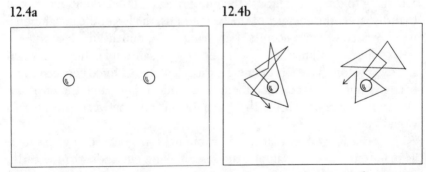

12.4a. *In classical physics we can think of two protons sitting quietly at rest.*
12.4b. *In quantum physics, the two protons are constantly driven by the wild dance of the quantum.*

Wonderful! Now let us try to follow the exact same reasoning and calculate the quantum correction to the gravitational force between two protons at rest. What do we get? We get hit in the face by the sheer nonsense of the result. Instead of being tiny, the quantum correction turns out to be infinitely huge!

THEORY CONTRADICTS FACT

Why does a calculation that works so well with the electric force fail so miserably with gravity?

As the protons dance, they possess energy associated with their wild rocking and rolling. The energy fluctuates. Most of the time the energy is small, but once in a while it can be enormous.

Hey, wait a minute, you say, I know enough physics to know that energy is conserved. Energy can't appear out of thin air! Ah, yes, but it is a weirdness of the quantum world that energy can appear out of, and disappear into, thin air, although only for a short time. We are talking about times so short as to be virtually inconceivable by the human mind. The larger the amount of energy that popped out of thin air, the shorter the time. A crude and decidedly nonquantum analogy is fluctuation in the weather. In winter we may suddenly have a few warm days in succession. The warmer those unseasonable fluctuations, the shorter they last.

To calculate the quantum correction to the force between two protons, we have to add up the effects of all possible quantum fluctuations according to the probability of their occurring and their durations. Recall that the larger the energies of the two protons, the larger the gravitational force between them becomes. Now comes the punch line.

For short bursts of time governed by quantum fluctuations, when the two protons have lots of energy, the gravitational force between them becomes huge. In fact, since there is no limit to how large the quantum energy fluctuations can be, the gravitational force between two protons would become infinitely huge.

But it is a fact that the gravitational force between two protons, as manifested in, for example, the force pulling our bodies to the earth, is not infinitely huge.

Theory contradicts fact.

Therefore, theory is sheer nonsense. Given our present understanding of quantum physics, the existing theory of gravity can't be right.

The mystery is that the theory of gravity works perfectly as a classical theory. Only when we force the quantum on it does the theory fall flat on its face.

This discussion underlines the sharp contrast between the gravitational force and the electric force. The electric force does not become strong with energy, and thus the quantum correction remains tiny despite the large energy fluctuations. In the final analysis, the trouble with gravity is that it is sensitive to mass. Mass is equivalent to energy according to Einstein, but energy as a physical concept can be arbitrarily large. In contrast, electromagnetism is sensitive to charge, which, as far as we know, has a fixed value. The theory of electromagnetism is renormalizable—that is, it can be combined with the quantum principle—but the theory of gravity is not.

GRAVITY GOES BERSERK

In the preceding discussion, for each quantum energy fluctuation we can calculate the correction to the gravitational force between the two protons. The larger the energy fluctuation, the larger the correction. Clearly, a relevant question to ask is the following: How large is the energy fluctuation when the corresponding correction becomes comparable to the classical value? The answer: an energy of about 10^{19} GeV. This energy is known as the Planck energy, in honor of Max Planck, one of the founding fathers of quantum physics. At the Planck energy, gravity goes berserk: The quantum correction starts to overwhelm the classical value. Again, the enormity of the Planck energy clearly just reflects how weak gravity is.

(You may have noticed that the Planck energy is within a factor of ten the energy of 10^{18} GeV at which gravity overwhelms electromagnetism. The physical origins of the two energies are different, however. The Planck energy is quantum in origin and refers only to one force, while the energy of 10^{18} GeV emerges from comparing two forces classically.)

To use the classical theory of gravity, physicists have to confess ignorance of the physics above the Planck energy. That is perfectly okay. Physics is an empirical science, and the Planck energy is far beyond the realm that has been explored by experiments. When I taught the subject, to underscore this point I would write on the blackboard in big letters, "Ignorance is no crime." Why should we trust the known laws of gravity, and for that matter, quantum physics, at the Planck energy? Absolutely no reason. (Figure 12.5.)

Thus, in calculating the quantum correction to the gravitational force between two protons, we may arguably ignore the contributions coming from the fluctuations at energies beyond the Planck scale. Physicists say that they cut off the calculation. The quantum correction is then reduced from infinitely huge to uncomfortably large. This does not represent some kind of legerdemain, but a confession of ignorance. By cutting off the calculation, physicists don't understand any better why the gravitational force is not infinitely huge; the calculation has merely been raised from the level of sheer nonsense to that of total nonunderstanding.

That the Planck energy turned out to be so enormous offers some consolation. Had the Planck energy turned out to fall within a realm that has already been explored by experiments, we would be forced to regard the theory of gravity as sheer nonsense.

I should perhaps note that in the phrase "a realm that has already been explored by experiments" I allow for both direct and indirect explo-

12.5. *A physicist cutting off his calculation.*

ration. Thus, many physicists, including myself, believe in the grand unified theory governing physics at an energy scale of 10^{15} GeV. While this regime has not been explored directly, it has been explored indirectly in that experimenters have measured certain quantities whose predicted values depend on calculations involving the physics at 10^{15} GeV. The point is that while our present understanding of the physics at 10^{15} GeV may turn out to be embarrassingly incorrect, there is no reason to suspect at present that it is incorrect. In contrast, we know for sure that our present understanding of physics must fail at 10^{19} GeV.

Remember our discussion of characteristic energies in Chapter 6 as we looked back toward the big bang? The characteristic energy of particle physics is around 1 GeV, of nuclear physics 1 MeV (=one thousandth of a GeV), and of atomic physics 10 eV (=one hundred-thousandth of an MeV). If we were studying physics at energies of 10 eV or so, we wouldn't know about the effects of particle physics at all.

What we are learning here is that the characteristic energy of gravity is 10^{19} GeV. We wouldn't have known about gravity at all if it didn't so happen that the universe contains a humongous number of particles and that a huge number of these particles had come together into chunks called earth, apple, and humans. This is another way of saying that if we watch two protons interacting, the gravitational force between them is totally drowned out by the other three forces. Gravity manifests itself only because we have big lumps of matter around.

Gravity is a mysterious interloper from an energy scale far beyond our experience. By happenstance we have made its acquaintance.

FORETELLING ITS OWN DOOM

That a physical theory can predict when it must fail and will fail is to me one of the remarkable feature of physics. Of course, physicists have always understood that the so-called phenomenological theories—theories constructed to account for specific phenomena—have limited domains of validity. For instance, the theory of elasticity assumes that the force required to stretch an elastic spring is proportional to the amount by which the spring is stretched. Obviously, the theory fails when the force is so large as to break the spring. In contrast, fundamental theories, such as Newtonian mechanics, often looked as if they would be universally applicable. That Newtonian mechanics fails totally in the submicroscopic world was not anticipated. Viewed in this perspective, the fact that existing physics can predict its own demise represents an astonishing novelty.

Incidentally, I mentioned earlier that until about 1971 the quantum theory of the weak force was also nonrenormalizable. It predicted it can't go beyond 300 GeV or so. Indeed, the new theory of the weak force that emerged after 1971 indicated that the weak force as we knew it would be modified at about 100 GeV. Subsequent experiments have shown that this is indeed the case. This episode gives physicists confidence in their conclusion that gravity must be modified at 10^{19} GeV. While musing on this amazing ability of physics to proclaim its own limits, I am driven sometimes to wonder wistfully what it would be like if political and economical theories could also proclaim their limits of applicability. Why should a political-economical theory constructed in response to conditions in industrial Victorian England be *necessarily* applicable to twentieth- and twenty-first-century societies?

ERA OF DARK IGNORANCE

One day not long ago, I sat in a pretty garden shooting the breeze with some friends. One of them asked me, "Tell me, what actually happened right at the big bang?" When I told him that I didn't know, he looked at me with distinct skepticism. "Are you sure you are really a physicist, as you said you were?" his look implied.

The limit to present-day physics sets a limit to our discussion of the early universe. As we journey toward the big bang, we eventually reach an era of dark ignorance, covering the time when the universe was hotter than the Planck temperature. (The Planck temperature, as you might guess, is the temperature at which particles have characteristic energy of the order of the Planck energy.) Physicists have *not the foggiest idea* of what happened during the era of dark ignorance.

As a working physicist, I react with displeasure to popular expositions on physics and on the cosmos that speak of the Moment of Creation and even of How the Universe Was Created. The fact is that research physicists cannot and do not speak of the very moment of the big bang. Remember what Gamow said back in Chapter 6. Hey, you understand nuclear physics. Fine; then you can discuss the universe at nuclear temperatures. You think you understand grand unified physics—well, you are not so sure . . . All right, you can speculate about the universe at grand unified temperatures. What? You don't have the foggiest notion about Planck scale physics! Well then, how are you going to talk about the universe at the Planck temperature?

Indeed, with our present understanding, we may as well say that the universe began when the era of dark ignorance ended, when the universe's temperature dropped below the Planck temperature. We could then define the big bang as the transition between the era of dark ignorance and the era of lucid knowledge.

We can say with some confidence that the era of dark ignorance ended some ten billion years ago, but it makes no sense to discuss how long the era lasted. Nevertheless, physicists often say that the era of dark ignorance ended some 10^{-44} seconds after the big bang. What the professionals mean by that is the following: Suppose we simply ignore quantum physics and use Einstein's theory of gravity. Then we do reach a beginning, *defined* as the time when the temperature of the universe reaches infinity. We can then ask, according to classical gravity, how long it would take the universe to cool down to the Planck temperature. The answer is 10^{-44} second. Since we know for a fact that classical gravity does not apply in that era, this number is not particularly meaningful. Incidentally, this discussion again points out that what is relevant for physics is not so much the time, but the temperature at which various processes occurred.

THE FATE OF THE UNIVERSE

After all this talk about the era of dark ignorance, I would like to say a few words about the fate of the universe. Ominous-sounding, isn't it? I feel I should capitalize the phrase. Poets and scientists, philosophers and theologians all have been fascinated by the question of how the universe will end. There is, of course, Robert Frost's opinion, on the tongue of every schoolboy and grown colorless from excessive repetition. ("Some say the world will end in fire, / Some say in ice./ From what I've tasted of desire / I hold with those who favor fire.") With our present understanding, the fate of the universe hinges simply on whether the universe contains enough matter per unit volume to halt the expansion, as discussed in Chapter 11.

THE COLD VOID OF FOREVER

If the universe contains little matter, then it will go on expanding forever. Soon, the starfires will go out, having exhausted their nuclear fuel. Matter pulls farther and farther apart. The dark nights of eternity become colder and colder. And thus will the universe end, with not even Eliot's whimper but with a long, long sigh that gradually dies out.

However, if the speculations of grand unified theory are correct, the universe will live bathed in the glory of light for a while longer. Recall that grand unified theory suggests that the proton does not live forever. Suppose, to be definite, that the proton lives on the average for 10^{32} years. Every year, each proton in a star has one chance in 10^{32} of kicking off; in effect going out with a burst of light. Normally, the light put out by the dying protons is completely overwhelmed by the light put out by the nuclear reactions inside stars. But as stars exhaust their nuclear fuel, the death glow of the protons becomes ever more apparent. The stars continue to glow faintly; the sun, for instance, will glow some 10^{18} times dimmer than in its nuclear burning heyday. After 10^{32} years, the lights finally go out. The universe steadily expands into the cold void of eternity.

PURGED OF ITS IMPERFECTIONS

On the other hand, if the universe contains lots of matter, then the gravitational pull of the matter will eventually rein in the expansion of the universe. The expansion gradually slows until it finally stops. Then the universe starts to contract at an ever-faster rate.

As the universe contracts, its temperature rises accordingly. The universe relives its adolescence, as described in Chapters 6 and 7, but in reverse. Soon planets, stars, and galaxies are crushed together and vaporized into a broth of electrons and atomic nuclei. The violence and mayhem get out of hand. Rampaging protons and neutrons break the nuclei into their constituent protons and neutrons. The protons and neutrons, in turn, smash each other into quarks. The universe now looks the way it looked shortly after it started. Finally, when the temperature reaches the Planck temperature, we once again enter the era of dark ignorance, and given our present understanding, neither I nor anyone else can tell you what will happen afterward.

OUT OF DEATH, REBIRTH?

One intriguing possibility is that when the contraction reaches a certain point, the universe will rebound and expand. From apparent death, the universe is born again. My friends who are not physicists almost invariably prefer a universe with the possibility of rebirth to a universe expiring into the cold void of forever. There is something enormously appealing about a life after death for the universe. The universe, purged of all its imperfections in a fiery catharsis, rises again for another day, another try. In this scenario, the universe has neither beginning nor end. It goes on in perpetuity, cycling from birth to death and then to birth again.

As a physicist, I must emphasize again that we do not know anything about the physics above the Planck temperature and thus have no way of knowing whether the universe will rebound.

ENTERING THE DARK ERA

Incidentally, recall that if the universe once went through an inflationary phase during which it expanded rapidly, then the mass density of the universe is exactly equal to that special value that divides a universe that expands forever from a universe that eventually contracts. You may be wondering what would happen to a universe poised precisely on the dividing line. As a child, I wondered which way a tilting bus just teetering on the edge would decide to fall. (Figure 12.1.)

What does the universe decide to do? To expand forever or not to expand forever? (See Figure 12.6.) The answer is that such a universe will expand forever. If you believe that we live in an inflationary universe, as is currently fashionable, then our universe has but one life to live.

12.6. *To expand forever or not to expand forever? In a sense, that's not the question. The more interesting questions are how much dark matter the universe contains and what is the nature of the dark matter.*

13

The Music of Strings

A CABARET SHOW

In the summer of 1984, my wife and I drove across the country with our children, stopping at various physics centers along the way, much like migratory wildebeests seeking out watering holes. At Aspen, Colorado, we stopped for a month. There, amid the splendors of the Rocky Mountains, physicists gather every summer to discuss the latest sense and nonsense. The atmosphere at Aspen always has been rather relaxed: The physics chitchat is mixed in with volleyball and picnics and hikes and music and the local scene, such as it is.

One warm summer's eve, as part of the fun and frivolity, the physicists took over the bar at the venerable Hotel Jerome to stage a variety show. In one of the skits, a physicist came out ranting and raving that at long last he understood the secrets of gravity. He now had a theory of the whole world, including gravity. Before he could explain what it was, however, two men dressed in white coats came and dragged him away.

The skit is a familiar one, and the sight of a mad physicist being dragged off to an asylum can always be counted on to provoke a wave of titters among a physics audience. Sitting in the audience, I felt that there was a certain Gary Larson flavor to the whole scene. But that summer, the physicist playing the mad physicist was perfectly serious about having a theory of the whole world.

Some days earlier, when I first arrived at the Aspen Center for Physics, I ran straight into John Schwarz, who played the mad physicist, and his collaborator Mike Green. They were both excited. Mike Green and I had been postdoctoral fellows together, and I know well how he talks when he gets excited. Green and Schwarz had been working on something called the string theory, later to become the superstring theory, on and off for a decade, and now finally they had a version of the theory that works.

NOT INFECTED WITH ANOMALY

When physicists write down a theory of the world, they must check whether the theory can be combined with the quantum principle. In the 1960s, physicists discovered that even though some theories may look perfectly sensible at first, they can harbor a mathematical inconsistency called the *anomaly*. If your favorite theory suffers from the anomaly, then it cannot be combined with the quantum principle and you can kiss it good-bye. To check whether a theory is infected with the anomaly, physicists developed what amounts to a clinical test. You look at your theory and calculate a number according to some formula. That number determines the degree to which the theory is infected with the anomaly. If you get zero, the theory is not infected. If you don't get zero, it goes into the trashbasket.

Superstring theory is notorious for being afflicted with the anomaly, and that partly accounts for the general lack of interest in the superstring theory till that summer. Green and Schwarz had been computing the degree to which various versions of the superstring theory were afflicted with the anomaly. Mmm, not zero. No good. Try another one. Finally they found a version in which the number, remarkably enough, happens to be zero. Hence their enormous excitement when I ran into them. They showed me on the blackboard how the numbers precisely cancel. I was particularly interested, since throughout my career I had worked on the anomaly from time to time. I was astonished by the almost magical way in which the anomaly in the Green and Schwarz version of the superstring theory cancels to zero. John Schwarz likes to refer to this cancellation as the first miracle of superstring theory.

COMING FROM LEFT FIELD

To understand what all the excitement about superstring theory is, we have to go back to all those difficulties bedeviling the theory of gravity. For a good part of this century, many physicists have tried to unify gravity with the other forces, to marry gravity to the quantum, and to sneak a peek into the era of dark ignorance, but they all failed. As is often the case in physics, progress did not come from those banging their heads against the wall and moaning about gravity, but from an entirely different crowd. The string and the superstring theories were originally invented to describe the behavior of the strongly interacting particles of the subnuclear world—and failed miserably. It was not until 1974, when John Schwarz and the late French physicist Joel Scherk suggested that the theory was being applied to the wrong set of phenomena, that it should be used to describe gravity instead. And now in 1984 Green and Schwarz were saying that the superstring theory, in those versions certified to be free from anomaly, is in fact a sensible theory of gravity.

STRING AND SUPERSTRING

Over the years, the language of physics has been refined steadily, arriving in our times at the enormous sophistication of the quantum field theory. Sophisticated though it is, the quantum field theory is ultimately still based on the simple intuitive notion that material objects can be divided into particles and that these particles are like tiny balls that can be treated mathematically as points, a notion that has been with physics almost from the beginning and that most physicists and I personally find rather appealing.

It is in fact astonishing that physics has managed to describe Nature so well for so long with point particles. When Democritus talked about atoms, when Newton talked about corpuscles, they did not insist on atoms and corpuscles being strict mathematical points, as far as I know. After the discovery of atoms, it soon became clear that atoms were not points but were made of smaller entities. But thus far, all the evidence indicates that the particles fundamental physicists deal with—quarks, electrons, photons, and the like—are mathematical points within the accuracy of the experiments. For instance, experimenters have determined that the electron can't be bigger than about 10^{-18} centimeter.

In contrast, the string theory, which begat the superstring theory, is based on the notion that fundamental particles are actually tiny bits of strings. Indeed, if the bit of string is much shorter than the resolution of our detection instruments, it will look like a mathematical point. What we thought were point particles flying around in space and time are actually itsy-bitsy pieces of strings flying around, according to string theory enthusiasts. (These tiny strings are not to be confused with the cosmic strings we discussed in Chapter 9.)

As the bits of string move around, they wriggle like tiny worms. The remarkable feature of the string theory is that by wriggling in different ways, the string can appear to us as different particles. In fact, the string theory contains infinitely many different particles, since a piece of string can wriggle in infinitely many different ways.

UNIFICATION OF GRAVITY

How does the string theory unify gravity with the other forces?

Recall from Chapter 3 that Einstein's theory of gravity specifies completely the properties of the graviton, the fundamental particle of gravity. The graviton is massless, spins twice as fast as the photon, and so on. Remarkably, the converse also is true: If a theory contains a particle that has precisely the same properties as the graviton, then that theory contains Einstein's theory.

As it happens, one of the infinitely many particles contained in the string theory has precisely the properties of the graviton. Thus the theory automatically contains Einstein's theory of gravity. The fact that the theory contains the graviton is hailed by the faithful as the second miracle of the superstring theory.

Gravity is unified with the electromagnetic interaction because another one of the infinitely many particles has precisely the properties of the photon. A theorem analogous to the one I cited above states that in this case the theory contains the electromagnetic interaction. Similarly, it turns out that the string theory contains the strong and the weak interactions as well.

The theory does not so much unify gravity with the other three forces as contain gravity and the other three forces.

Gee, you may say, that's awfully simple. I could have thought of it. Indeed you might have. The basic conceptual idea is almost embarrassingly simple. In contrast, the mathematical formulation is horrendously complicated. Part of this is due to our lack of familiarity with strings. For centuries, physicists played with point particles. To deal with strings, we suddenly have to extend our entire formulation of physics. What is worse, the intuition we have built by playing with point particles is by and large irrelevant.

What is so horrendously complicated about vibrating strings? Even introductory books on physics often talk about vibrating violin strings. Well, first of all, these strings are vibrating at the speed of light. Second, one has to keep track of the infinitely many ways in which the string can vibrate. Third, when one quantizes the theory, as one must to describe our quantum world, particles with totally unacceptable properties—particles that physicists have playfully dubbed ghost particles—threaten to appear. It takes an exquisitely elaborate analysis to show the ghost particles to be merely mathematical fictions. (To be fair, I must mention that when the quantum field theory was first invented, it also struck physicists as horribly complicated. With the passage of time, the horror gradually subsided—partly because physicists became familiar with the quantum field theory, and partly because they developed new ways of looking at the quantum field theory that made it seem a lot simpler. Many physicists believe that the same developments will happen to the string theory.)

STRINGS GO SUPER

Unfortunately, the basic theory of strings as I have described it, the sort of theory you and your Uncle John might have written down, turns out to be glaringly inadequate: It does not even contain the electron! That is because you and your Uncle John, being perfectly reasonable and sensible people, would have described the string vibrations in a reasonable and sensible way, something like the following: As one tenth of a second has elapsed, this point on the string has moved west by 0.27 inch, this other point has moved west by northwest by 0.18 inch, and so on until the movement of every point on the string has been specified. You would have described how the string has moved from one particular instant in time to an instant one tenth of a second later. To include the electron, physicists had to extend this description. They say that in one tenth of a second this point on the string has moved in some direction by 0.27 inch and ψ, where ψ represents a weird sort of number invented by a mathematician named Grassman. Physicists represent Grassman numbers by a Greek symbol such as ψ.

What is a Grassman number? In a nutshell, a Grassman number

is a number such that when you multiply any Grassman number by itself, you get zero. Thus, $\psi \times \psi = 0$. Whaaat? Grassman numbers are weird. How can you possibly multiply a number by itself and get zero? Only a mathematician would think of such a thing.

A string theory in which the strings are allowed to move by amounts described by Grassman numbers as well as by ordinary numbers is known as a superstring theory.

I can't possibly give you in less than thirty pages a detailed explanation of how physicists use Grassman numbers, and I am sure you would be bored stiff. I mention Grassman numbers only to show you one weird aspect of the theory. Once the theory is detached from its moorings on ordinary numbers, physicists have a hard time visualizing it as describing vibrating strings. If you think string theory is abstruse, superstring theory is even more so.

Not impressed! you say. What is so profound about going from points to strings? Nothing, physicists would have to concede if hard-pressed. Then why did it take so long for physicists to move from a theory of points to a theory of strings? The answer is that physicists had been making such tremendous progress with theories based on point particles that there was not a strong motivation to go to theories based on strings. Actually, over the years there had been sporadic attempts to write down theories not based on point particles, but physicists had generally recoiled in horror at the attendant complications. One must pay tribute to pioneers of the string theory such as John Schwarz and Michael Green for their persistence.

BEYOND STRINGS?

Heck, I can make a new theory, too, you exclaim. If you guys just moved on from theories of points to theories of strings or curved lines, why stop? As mathematical objects, points are zero-dimensional, lines are one-dimensional, surfaces are two-dimensional, and ball-like blobs are three-dimensional. Why not continue the natural progression and move on to theories of surfaces and then theories of blobs? Just as a teeny bit of string can look like a point particle, so a teeny bit of surface or membrane can look like a teeny bit of string. Why not a membrane theory and a supermembrane theory, a blob theory and a superblob theory?

People have tried to write down membrane theories and blob theories, but these theories are so complicated that physicists couldn't get past square one in analyzing them. But surely, Whoever designed the universe couldn't care less whether humans in the late twentieth century should have enough mathematical pizzazz to solve the puzzle. Of course, the string theory is presumably as good an approximation of the membrane theory as the point particle theory is an approximation of the string theory. String theory enthusiasts would concede that theirs might not be the ultimate theory either but merely a better theory than the point particle theory. Still, many physicists, and I for one, are bothered that by moving away from points, physicists might be opening up an infinite regression of Russian dolls.

AN EXTRAVAGANT UNIFICATION

Gravity is finally unified with the three other fundamental forces, in the sense that the superstring theory contains the graviton and the fundamental particles associated with the other forces, such as the photon. But the theory also contains an infinite number of other particles. These particles all have enormous masses, at least 10^{19} times the mass of the proton. According to the terminology introduced in the preceding chapter, these particles have masses of the order of the Planck energy and beyond. At "ordinary" energies—that is, at energies far below the Planck energy—they are too massive to participate. They come in only when the energy available exceeds their mass.

One might say that gravity is unified with the strong, the electromagnetic, and the weak forces, but at the same time also with an infinity of other forces that do not show up at ordinary energies. It is what might be called an extravagant unification. Gravity has been persuaded to join the dance, but only at the cost of hiring an infinite number of extras.

How does string theory manage to marry gravity to the quantum? Recall that when physicists tried to turn Einstein's theory of gravity into a quantum theory, the quantum fluctuations with energy above the Planck energy produce an infinitely large correction to the gravitational force between two objects. How does the string theory get around this difficulty? Roughly speaking, when the energy of the quantum fluctuations exceeds the Planck scale, all those "extras" in the drama of physics, the whole infinitude of them, come to the fore and sing their song. Hey hey, they sing, at such high energy, we can dance to the tune of the quantum and generate quantum fluctuations like everybody else. These fluctuations also produce an infinitely large correction to the gravitational force, which, miracle of miracles, cancels the unacceptable correction produced in Einstein's theory.

To put it in another way, Einstein's theory of gravity is merely an approximation of a piece of the string theory for energies below the Planck mass. For energies above the Planck scale, it's another ball game, and calculations of quantum fluctuations using Einstein's theory do not tell the whole story.

MULTIDIMENSIONAL UNIVERSE

An astonishing feature of the superstring theory is that it can be formulated only if space is nine-dimensional. Upon first hearing this, you might want to chuck the theory out the window, and that was indeed the initial reaction of many physicists. Space is evidently three-dimensional to any sensible person. Einstein had unified space with time and described the physical world as four-dimensional: the three familiar dimensions of space and the dimension of time. But to Einstein, space remains three-dimensional, as any ordinary person would have thought. The superstring theory, however, makes sense only if there are six extra dimensions to space.

How could we possibly have missed these extra dimensions? Easily, if these extra dimensions are tiny. Consider a creature constrained to live on the surface of a tube. The space inhabited by the creature, the surface of the tube, is really two-dimensional. Suppose the radius of the tube is much smaller than the smallest distance that can be perceived by the creature. To the creature, space would appear to be one-dimensional; since the creature can move only along the tube.

In other words, we can mistake a very thin tube for a line. On closer inspection, every "point" on the "line" actually turns out to be a circle. Thus it is quite possible that every point in the familiar three-dimensional space in which we move would, on closer inspection, also turn out to be a circle. If the radii of the circles are much smaller than the smallest distance we can measure, we would see the circles as points, and we would be misled into thinking that we are living in a three-dimensional space instead of a four-dimensional space. Don't worry if you can't visualize such a strange space. Neither can the myopic creature living on the long, hollow tube visualize that his space actually is two-dimensional.

We can keep going. Perhaps on closer inspection, each point in our three-dimensional space would turn out to be a two-dimensional surface, such as the surface of a sphere. In that case, our beloved three-dimensional space would actually be five-dimensional, and so on. Physicists call the three-dimensional space that we know external space and the space hidden from view because of our myopia internal space. The actual dimensions of space are then the dimensions of internal space plus three.

Experimentally, there is no evidence whatsoever of an internal space. Experimenters have looked and said that the internal space, if it exists at all, must be a few hundred times or so smaller than the size of the proton. Now, that is absolutely teeny compared to the human scale of things, and we must congratulate the experimenters for their heroic efforts in looking down to such tiny scales. But while experimenters work hard, theorists are free to believe in an internal space. They merely have to snap their fingers and chant in unison that the internal space is much smaller than what the experimenters can measure. In fact, string theorists believe the size of the internal space to be 10^{18} times smaller than the size of the proton. There is no hope for experiments to be able to detect a space of this size within the foreseeable future.

ELECTROMAGNETISM OUT OF GRAVITY

The idea that space has hidden internal dimensions is in fact recovered from the dustbins of history. In 1919, just four years after Einstein proposed his theory of gravity, the Polish mathematician and linguist Theodor Kaluza suggested that space may be four-dimensional. The Swedish physicist Oskar Klein then developed Kaluza's work into what is known as the Kaluza-Klein theory. By the way, when Einstein learned of these ideas, he was astonished. He wrote to Kaluza that the notion of four-dimensional space had never occurred to him.

Any sensible person might react to Kaluza's suggestion by asking what good an extra hidden dimension of space is. You can assert that any sort of fancy chimera exists just by insisting that it is too small to be seen.

To be sure, for physicists to have paid any attention at all to Kaluza and Klein, they must have found a wonderful benefit in hidden dimensions. They discovered that the Kaluza-Klein theory unifies gravity with electromagnetism.

Consider the simplest possibility: that the internal space is a circle and that the world is actually five-dimensional, with four dimensions for space and one dimension for time. Suppose, Kaluza and Klein said, that there is only gravity in the world. How would the inhabitants of this world, too myopic to see that what they call points are actually circles, perceive the gravitational force? To Kaluza and Klein's utter surprise, they found that the inhabitants would feel two types of forces, which they could interpret as a gravitational force and an electromagnetic force! In Kaluza and Klein's theory, electromagnetism comes out of gravity!

We can understand this stunning discovery roughly as follows. A force in a three-dimensional space can pull in three different directions; after all, that is what we mean by saying that space is three-dimensional. In the four-dimensional space of the Kaluza-Klein theory, gravity can pull in four different directions. To us, the myopic inhabitants, a gravitational pull in the three directions corresponding to the three directions we know appears as just a gravitational pull. But what about a gravitational pull in the fourth direction, the direction that we are too myopic to see? We would construe that as another force and call it electromagnetic.

To many physicists, the emergence of electromagnetism from gravity can only be described as "mind-blowing." Nevertheless, the Kaluza-Klein theory faded away from the collective consciousness of physicists as it became clear that there was more to the world than gravity and electromagnetism. The strong and the weak forces had to be invented to account for a whole host of new phenomena, and the strong and the weak forces, as they were understood, just did not fit into the Kaluza-Klein framework. When I studied physics, I never heard of the Kaluza-Klein theory. The major textbooks on gravity in the 1970s never mentioned it.

FORCES FROM GEOMETRY

Even while the Kaluza-Klein theory was in disrepute, a few physicists went back to the theory periodically. Kaluza and Klein took the internal space to be a one-dimensional circle. What if the internal space is multidimensional? For example, the internal space could be a two-dimensional sphere. (See Figure 13.1a.)

These physicists found that the more dimensions the internal space has, the larger the variety of forces that emerge. If the internal space is a circle, the electromagnetic force emerges. If the internal space is a sphere, a collection of forces containing the electromagnetic force emerges. This is perhaps not surprising in light of the argument given above: There are more directions on the sphere in which forces can pull than on the circle.

How do these forces depend on the symmetry of the internal space? For example, a two-dimensional internal space could be a sphere or a torus. The torus is symmetric under rotations around the axis indicated by the dotted line in Figure 13.1b, while the sphere is symmetric under rotations around any axis. It turns out that if the internal space is a sphere, more symmetries relate the forces that emerge than would be the case if the internal space is a torus. Indeed, if the internal space has no symmetry at all, as indicated by the space in Figure 13.1c, then no force emerges. The number of forces that emerge depends on how symmetric the internal space is. Thus more forces emerge from the sphere, which is more symmetrical than the circle.

It is so incredibly neat! a Kaluza-Klein aficionado would exclaim.

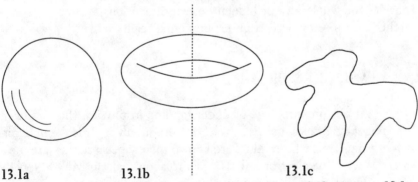

13.1a   13.1b   13.1c

Possible choices for the internal space. 13.1a. A sphere. 13.1b. A torus. 13.1c. A surface of no particular symmetry. For the meaning of the dotted line in 13.1b, see the text. In the figure, the space is by necessity represented as two-dimensional surfaces. In fact, physicists typically think of higher-dimensional spaces.

The geometrical symmetry of the internal space imprints itself on the symmetry of the forces we see in the external world.

Physicists have gone into ecstasies over the symmetries of the fundamental forces. Meanwhile, mathematicians have long exulted over the symmetries of geometrical shapes. In the Kaluza-Klein theory, the symmetries of the mathematicians come together with the symmetries of the physicists. This metamorphosis of geometrical symmetry into physical symmetry is extremely beautiful to watch, but unfortunately it can be appreciated fully only when cloaked in the splendor of mathematics.

By the 1970s, physicists had reached a new understanding of the strong and the weak forces. With this new understanding they managed to unify the strong, the electromagnetic, and the weak forces into one. It was then realized that this grand unified force could emerge from the Kaluza-Klein theory as a piece of gravity. At that point, the Kaluza-Klein theory came roaring back. Physicists tried different internal spaces. I for one was convinced that the internal space has to be a sphere, that most aesthetically satisfying of geometrical shapes.

I should emphasize that the idea of space having more than three dimensions is by no means universally accepted. Why is the internal space so small while the external space—namely, the whole universe—is so huge? Different dimensions of space presumably started out on the same footing, but for some unaccountable reason, some of them have stretched out, while others have shriveled up to almost nothing. Some physicists consider the whole idea of hidden unseen dimensions as simply too far-fetched and too removed from experimental reality.

A BROKEN PROMISE

At our present stage of understanding in physics, the masses of fundamental particles such as the electron are simply regarded as undetermined parameters whose values are known merely because they are measured. Why is the electron's mass equal to about half an MeV? Nobody knows. The Kaluza-Klein theory, however, asserts that the masses of fundamental particles are determined by the size of the internal space.

Physicists were excited by this bold promise of the Kaluza-Klein theory. Unfortunately, a quick calculation immediately threw cold water over their excitement: The electron's mass came out to be about the

Planck mass, 10^{19} GeV. The theory is a dog! The answer, 10^{19} GeV—that is, 10^{21} MeV—is not even in the same ballpark as half an MeV. In the Kaluza-Klein theory as we know it, the smaller the internal space, the larger the masses of particles like the electron. Since we can't make the internal space large, we can't make the electron mass small.

In the late 1970s and early 1980s, physicists tried a variety of ways of modifying Kaluza-Klein so the electron would come out to have a small mass. Nothing worked. In frustration, they were ready to give up on the theory.

It was more or less at this point that the superstring theory came on the scene. As we were saying before we launched into this long stretch of background on the Kaluza-Klein theory, the superstring theory makes sense only if space is nine-dimensional. With all this ongoing flirtation with the Kaluza-Klein theory, physicists took the news calmly. Far from chucking the theory out the window, they immediately absorbed the Kaluza-Klein idea into the superstring theory.

In what is sometimes known as the third miracle of the superstring theory, the electron comes out to be massless. Now, you might not think that this should count as a miracle, since the electron did not come out to have a mass of half an MeV. Well, compared to the Planck mass scale, half an MeV is for all intents and purposes equal to zero. Certainly it is a huge improvement over the Kaluza-Klein theory. Believers in the superstring theory are hoping that eventually, when the theory is well understood, some small effect now overlooked will correct this predicted value of the electron's mass from zero MeV to half an MeV. At this stage of the game, it is regarded as a minor detail.

TOO MUCH OF A GOOD THING

The Kaluza-Klein idea, however, could be implemented in the superstring theory only in a rather strange way. The difficulty is an embarrassment of riches. I mentioned that the string wriggling in different ways appears as different particles. Wriggling in one way, the string appears like a graviton; wriggling in another, it appears like a photon. Thus the theory contains the photon and the associated electromagnetic force. Similarly, it contains the strong and the weak forces. The theory already contains these known forces; it needs the forces that would come out of a Kaluza-Klein framework like it needs a hole in the head.

The string theorists are really in a bind. Physicists have striven to understand where the four known forces come from. Now suddenly there are too many forces: the forces from the wriggling string and the forces from the Kaluza-Klein framework. To avoid getting these Kaluza-Klein forces, string theorists are forced to choose internal spaces with no symmetry.

As I mentioned earlier, if the internal space is completely nonsymmetrical, forces do not emerge. By choosing a nonsymmetrical internal space, string theorists manage not to have too many forces: They keep only those from the wriggling string.

Traditionally, Kaluza-Klein aficionados have always considered symmetrical internal spaces. Why would anyone consider nonsymmetrical internal space? After all, the whole point of the Kaluza-Klein theory is to put in gravity and get out some other force besides gravity. In this sense, the implementation of the Kaluza-Klein idea in the superstring theory is against the original spirit of Kaluza and Klein. Some physicists, myself among them, are bothered by this subversion of the Kaluza-Klein theory. The beauty and advantage of going to a higher-dimensional space is that other forces emerge as pieces of gravity. But the superstring theory does not need these forces and so uses (according to some beholders, at least) the yuckiest possible internal space.

I for one am still hoping that the internal space will turn out to be a perfect sphere and that the original beauty of the Kaluza-Klein theory will be preserved. But my superstring friends dismiss this sort of sentiment as wishful thinking and ill-becoming nostalgia.

The complications of the superstring theory are due partly to the symmetrical internal space. With a nonsymmetrical internal space, physicists have great difficulties working out the consequences of the theory.

In Chapter 9 I mentioned how topological notions have entered physics and how by means of topology mathematicians can deduce properties about geometrical objects ordinary mortals cannot even visualize. In the Kaluza-Klein theory and the superstring theory, topological considerations really came to the fore. As long as the internal space is symmetrical like a sphere, its properties can be deduced readily. But when the internal space can barely be visualized, as in the superstring theory, then topological methods are needed to extract the answers to such physical questions as whether the electron's mass can be small compared to the Planck mass.

In the initial excitement, it was thought that the superstring theory could determine the internal space. Alas, that promise has proved to be illusory thus far. Many different types of internal spaces are possible, and to each of these spaces corresponds a different superstring theory. Physicists now have to rely on topology and other branches of higher mathematics such as algebraic geometry to help them classify and study these spaces.

HERE LIE DRAGONS

My lady friend is grumbling, "I still don't understand the superstring theory."

"Well, I know," I say with a sigh, "I am sorry. I didn't explain the superstring theory so much as describe it. Partly that's because the theory is both physically and mathematically so novel that physicists don't have a good understanding of what's going on. I didn't explain why the theory is not afflicted with the anomaly or why the electron mass comes out to zero. Nobody really understands all these *why*s. These miracles, and that's why they are jokingly referred to as miracles, just appeared out of the equations."

"That made me feel better. I was worried that I may have missed something." She appeared relieved.

"It doesn't make me feel any better. In writing a book on physics for an intelligent lay reader like you, my ideal is to try to explain as much as possible. But when I come to the latest, so much of it is based on the subtleties of quantum physics," I said rather ruefully.

"Well, that figures. I can at least picture a string wriggling. By wriggling in different ways, it can appear to us as different particles," she replied cheerfully.

"Right, that's the key point. The graviton is just one of an infinitude of wriggling forms. In a way, gravity has been demoted. That's one reason why some physicists are unhappy with the superstring theory. Somehow we feel that gravity should be special, being connected to space and time and all."

"Hmm, that kind of bothers me, too."

"But no doubt about it, superstring theory is by far the most significant development in our understanding of gravity since 1915. Ever since the arrival of quantum physics, gravity has pointed to its own demise. You know how mapmakers used to mark those regions they don't know anything about with a Latin inscription proclaiming, 'Beyond here lie dragons.' Gravity has marked the energy domain beyond the Planck energy with that inscription. Superstring theorists have now ventured into that region. Well, there are dragons! They are the wriggling string in its infinite number of forms."

MIRACLES AND DIFFICULTIES

With three miracles, the faith of the superstring enthusiasts was sealed. However, there are also difficulties and complications. At the moment, the physics community is divided over whether the superstring theory will turn out to be the ultimate theory of everything, as the enthusiasts proclaim. The theory is enormously complicated to develop, and the stage has not yet been reached when the issue can be settled decisively by experiments. An intrinsic difficulty is that while experiments can be conducted only at energies far below the Planck energy, the superstring theory is formulated expressly at that lofty energy scale. All those infinite numbers of extras introduced by the superstring theory— the supermassive particles representing all the different ways in which the string can wriggle—come out to sing their song only at the Planck energy and beyond.

RUNGS ON A LADDER

To understand better the differences in opinion over the superstring theory, let us discuss the energy scales of physics in more detail. These energy scales may be visualized as rungs on a ladder. On each rung physicists are trying to work out all the phenomena characterized by that energy scale. Consider, for example, an atomic physicist trying to calculate the properties of the sodium atom, which happens to have eleven electrons whizzing around a nucleus. As far as he is concerned, the sodium nucleus is just a tiny ball characterized by a few numbers such as its mass, charge, and magnetic moment. (A nucleus's magnetic moment tells us how it would behave in a magnetic field, just as its charge tells us how it would behave in an electric field.) He regards these numbers as givens, either directly measured or calculated, handed to him by a nuclear physicist. The nuclear physicist, on the other hand, tries to calculate the properties of the sodium nucleus knowing that it is made of eleven protons and twelve neutrons. She can't do her job, however, unless she is told about the properties of the proton and the neutron. Okay, if you tell me that the magnetic moments of the proton and of the neutron are such and such, then I can calculate the magnetic moment of the sodium nucleus. The particle physicist, in his turn, tries to calculate the magnetic moments of the proton and of the neutron.

Thus there is an orderly progression in our understanding of the physical world. On any given rung, physicists have to be given a bunch of numbers, sometimes known as parameters. As we move up the rungs of the ladder, we hope that the number of parameters needed at each rung diminishes. Thus, atomic physicists treat the magnetic moments of some several hundred-odd atomic nuclei all as parameters, while nuclear physicists regard the magnetic moments of only the proton and the neutron as parameters. (As far as magnetic moments are concerned, the story actually ends at the particle physics rung. The presently accepted particle theory fixes the magnetic moments of quarks with some definite values, while the magnetic moments of the proton and the neutron can be calculated in terms of the magnetic moments of the quarks.)

Even as physics explores ever higher energy scales, a myriad of phenomena remain to be understood at lower energy scales. For instance, the behaviors of various metallic substances continue to confound the expectations of physicists. However, the explanations of these phenomena surely do not lie in the physics of higher energy scales. The disintegration of a subnuclear particle in a zillionth of a second is hardly going to affect the behavior of metals.

Our understanding of the physical world thus can be quantified roughly by giving the number of parameters needed. The currently accepted particle theory contains eighteen parameters. (The precise number depends on how you count and what you regard as understood.) The goal of fundamental physics is to reduce the number of parameters down to an absolute minimum. Some physicists even entertain the hope that ultimately the number of parameters will be reduced to zero.

Where is the next rung on the ladder after particle physics?

Physicists can only guess. From atomic physics, with its characteristic energy scale of 10 eV, we had to climb up a hundred thousand times higher in energy to reach nuclear physics, with its characteristic energy scale of 1 MeV (that is, one million electron volts). From nuclear physics we had to climb up a thousand times in energy to reach particle physics, with its characteristic energy scale of 1 GeV (that is, one billion electron volts). At present, some aspects of physics up to an energy scale of 100 GeV have already been explored at high-energy accelerators, but nothing dramatically new has been seen. (I am using "new" in a strict sense. To qualify as new, the physics involved has to be such that it requires structural modifications to the presently verified theory.)

Will we see new physics at 1,000 GeV? Or only at a much higher energy scale?

Much rides on the question of where the next rung is. Lots of taxpayers' money is involved. The physicists who are building the next generation of high-energy accelerators would dearly like to know. Should they design the machine to explore physics at 1,000 GeV (known as 1 TeV), or should they shoot for 100 TeV?

All we know for sure is that there is a rung at the Planck scale, at 10^{19} GeV. Gravity tells us so. Furthermore, the grand unified theory suggests that the strong, electromagnetic, and weak interactions merge into one interaction at 10^{15} GeV, but not everyone believes in grand unification. Meanwhile, we have explored physics experimentally only at energy scales up to about 10^2 GeV.

An extreme view holds that that's it, boys and girls. There's not gonna be any dramatically new physics between 10^2 GeV and the grand unification scale. Grand unification enthusiasts call this enormous region "the desert," in which they say experimenters will find nothing but "barren sand." Experimenters and physicists involved in building new accelerators hate this view, naturally. Outraged, they point to the history of physics. In the past, whenever we went up in energy, we encountered dramatically new physics whose existence we had had no inkling of. Surely, "the desert" will bloom and prove to be a land of milk and honey.

PHILOSOPHICAL DIFFERENCE

Viewed against this background, the difference in opinion over superstring theory reflects a real philosophical difference among physicists on how physics should progress: by great leaps upward, or by arduously climbing up rungs of the ladder. The superstring believers hold that the true physics lies at the Planck scale and that we have essentially understood, at least in broad outline, all of physics up to the Planck scale. In their boundless faith, they proclaim that they have found the Ultimate Design. In contrast, the infidels are people of little faith. They want their beliefs anchored in hard facts, not in the soft elegance of higher mathematics. They fear that their colleagues, in their joyous leap toward the Panck scale, may instead have arrived at a never-never land of mathematical enchantment.

This clash in philosophy has always existed, between the idealists and the empiricists, between Einstein and the vast majority of theoretical physicists whose days are largely spent pondering over data and performing routine calculations.

Einstein once said, "I want to know how God created this world. I am not interested in this or that phenomenon. I want to know His thoughts; the rest are details." What arrogant idealism! What provocation! Hear the gnashing sound of anger coming from those physicists who are dedicating their lives to this or that phenomenon!

Of course, most of physics, from the study of stars to the study of metals and materials, is concerned precisely with this or that phenomenon, and rightly so. We are speaking here of what may be called fundamental physics, the striving to understand how the world is put together ultimately, and of the philosophical differences that exist within the fundamental physics community itself.

The old man is unrepentant and speaks again: "I am convinced that we can discover by means of purely mathematical construction the concepts and the laws . . . which furnish the key to the understanding of natural phenomena. Experience may suggest the appropriate mathematical concepts, but they most certainly cannot be deduced from it. . . . In a certain sense, therefore, I hold it true that pure thought can grasp reality, as the ancients dreamed." Pure thought! What hubris, but what glorious hubris! This is the dream that lures boys and girls into becoming real theoretical physicists! Whew, understanding the world through pure thought!

The skeptics are snickering. Yeah, the ancients dreamed, but they didn't get very far. Democritus thought matter was made of atoms, and in cultures from Islamic to Chinese there were sages who had similar ideas. But we would have known nothing about atoms without the experiments of the late nineteenth and early twentieth centuries. Mao felt that dialectical materialism (remember Gamow and his exam!) required subnuclear particles to be made of "stratons" (from the word "stratum"). But without the big accelerators built in the 1950s and 1960s we probably would never have found out about quarks and gluons.

What then was Einstein talking about? Did physics ever progress by pure thought? Well, the best example (and, some would argue, the only example, if it's an example at all) is provided by the old man's own theory of gravity. From the thought that a falling man feels no gravity flow the secrets of gravity. Of course, the thought itself had to be built on a fact, in this case Galileo's observation that all objects fall at the same rate. The point is that the theory was not constructed laboriously step by step, with experiments providing guidance at each turn. It was born whole. Never mind that the birth process was painful—Einstein had to struggle for years—the theory was born whole.

But Einstein's theory is the exception to the rule. The struggle to understand the weak interaction, for example, is more closely characteristic of the development of fundamental physics. From the discovery of radioactivity to the present, the theory of weak interaction was built brick by brick, with bricks often removed and even whole wings demolished when new experiments contradicted the theory.

Einstein's sentiment reflects his intoxication with his own success in discovering the theory of gravity. Many felt, however, that the discovery of the theory of gravity was the one exceptional case in which pure thought could take physics far. Three quarters of a century later, believers in the superstring theory are once more flying the banner of pure thought. The notion that fundamental particles are in reality wriggling strings was a pure thought, unmotivated by any experiment.

Particle physicists are deeply divided. While some revel in "pure thinking," others feel that Einstein's dream—if it is represented by the superstring theory—has turned out to be a nightmare. In practice, the decision of an individual physicist to work or not to work on the superstring theory is often based less on a grand philosophical outlook than on other factors, such as temperament, ability, prospects for career advancement, and, perhaps most importantly, sloth and inertia. It takes a great deal of energy to master a radically new development, and many physicists, even if only half skeptical of superstring theory, have opted to continue doing whatever they have been doing.

14

The Thinking Man and the Laughing God

THE MYSTERY OF GRAVITY

We have traveled far, through time and space, from the plague years of the late seventeenth century to the plague years of the late twentieth century, from the apple, to the old man's toy, to wriggling bits of string. From our detour through a universe opened up for us by the old man's insight as expressed in his toy, we have finally returned to gaze at the mystery of gravity.

Do we understand gravity?

Not really.

I have saved till last an awful paradox that points to a big hole in our understanding of gravity.

AN AGE WITHOUT PARADOX

My friends in fundamental physics and I often sit around lamenting that there are hardly any major paradoxes around these days. We are talking of major-league paradoxes, complete contradictions, total disagreements, in which theory and experiment are light-eons apart. At any given time, there are plenty of phenomena that physicists do not understand. But in almost all these cases, the consensus is that they *could* eventually be understood in terms of existing theories. Paradoxes arise when physicists, in their cockiness, feel certain that they understand a phenomenon. Then Nature surprises them.

The resolution of major paradoxes has often revolutionized physics. Classical mechanics and electromagnetism assert that the electrons in atoms should spiral into the atomic nucleus in a short time. Yet atoms—and, by extension, the world—are obviously stable. The resolution of this paradox contributed to the invention of quantum physics.

Not every age has its paradoxes. There have been periods in the history of physics when physicists were content to work out one phenomenon after another, exhilarated in their feeling that they had finally mastered Nature. During the last decades of the nineteenth century, physicists felt that all of Nature could be understood through assiduous calculations using the known laws of mechanics and electromagnetism. As the nineteenth century drew to a close, an eminent physicist of the time pronounced that the task of physics was done. It remained only to calculate the next few decimal places. Nature was soon to wipe the smugness off physicists' faces.

Since the late 1970s, many fundamental physicists began again to assume the smugness of their late-nineteenth-century predecessors. The consensus is that we finally have the theory of the strong force, of the electromagnetic force, and of the weak force. We may even have a grand unified theory of these three forces. To be sure, the petulance of gravity is irritating. We don't understand physics at the Planck energy, but not having a good understanding is not the same as being stared at by a monstrous paradox. The feeling is that all of Nature short of the Planck energy could be understood by careful and often tedious calculations using the established theories. Superstringers even believe that they can understand Nature above the Planck scale. Suddenly physics seems to have become a mature science, even staid, particularly for those who lived through the excitement of the 1970s.

If our age is without paradox, it is certainly not for lack of trying. Physicists are revolutionaries at heart and ache to poke holes in the existing order. Both theorists and experimentalists have been gunning for a major paradox. In their eagerness, physicists make mistakes. Periodically, experimenters announce discoveries that confound all theoretical expectations. All the thrill-seekers in the community get uncontrollably excited. But alas, in the past two decades these theory-busting discoveries have invariably turned out to be wrong. Experiments in fundamental physics would confirm the established theory with an almost boring regularity. Or at least, no confirmed experiment has proved to be inconsistent with the existing theory. Taking a cue from our politicians, some wits in the physics community have dubbed this the doctrine of plausible deniability: At present, it can be plausibly denied that our existing theoretical framework is contradicted by any confirmed experiment.

Physicists have been hungry for a big, juicy, scandalous paradox to bite into.

LONGING FOR THE UNCHANGING

Well, there is a paradox, although it wasn't even widely recognized as a paradox until recent years.

The greatest paradox of our time brings us full circle back to the young Einstein's daydream that a falling man feels no gravity, a deep truth that was to be verified so playfully later by the old man's toy. As we saw, that happy thought led to the equivalence principle, which led in turn to curved space and warped time and a dramatic new theory of gravity. In 1917, Einstein was poised at the threshold of discovering the dynamics of the entire universe.

Alas, it was not to be. As I mentioned in Chapter 4, Einstein was wedded to the notion of an unchanging, eternal universe. When he started to study the universe, he first checked to see if his own theory of gravity would allow a static universe. When he saw that it didn't, he was upset and proceeded to muck around with the theory. (Figure 14.1.)

To understand what Einstein did, let us look at his famous equation of gravity. This theory has a deserved reputation of being difficult to master, but the bottom-line equation of the theory is not hard to understand:

curvature of space and time = distribution of mass and energy

This equation tells us that something is equal to something else. Written out in mathematical form, Einstein's equation says

$$R_{\mu\nu} - \tfrac{1}{2}g_{\mu\nu}R = -T_{\mu\nu}$$

The mess of symbols $R_{\mu\nu} - \tfrac{1}{2}g_{\mu\nu}R$ on the left is just the physicists' way of saying "curvature of space and time," while the symbols $T_{\mu\nu}$ on the right are their way of saying "distribution of mass and energy."

Okay, if you want to know what the curvature of the universe is,

14.1. *Einstein attempting to hold back the expansion of the universe.*

you have got to tell me how mass and energy are distributed in the universe. Once you tell me that, I just plug it into the equation and obtain the curvature of the universe. In particular, if you give me a uniform distribution of mass and energy, I would obtain a curvature of space and time that describes an expanding universe.

But no, Einstein didn't do that! Instead, he wanted an eternal universe so badly that he changed his equation to

$$R_{\mu\nu} - \tfrac{1}{2}g_{\mu\nu}R + \Lambda g_{\mu\nu} = -T_{\mu\nu}$$

He added an extra term, $\Lambda g\mu\nu$, now known to physicists as the cosmological constant term. The Greek letter Λ (lambda) represents the strength of the cosmological constant. Crudely speaking, this term can be thought of as describing the stretching of space and time. The modified equation says

curvature of space and time + stretch of space and time = distribution of mass and energy

By carefully adjusting the strength of the cosmological constant, Einstein was able to obtain a static universe.

Quite aside from the fact that the universe is not static, Einstein's move was rather distasteful. The universe stays static only if the effect of the cosmological constant exactly balances the effect of matter in the universe. Were there just a tiny bit less or a tiny bit more matter, the universe would start expanding or contracting. Einstein's solution is analogous to balancing a pencil on its point: Competing effects have to cancel perfectly for the universe to be balanced.

Well, as we know, the universe does expand, and Einstein was wrong. Sorry, gang, forget the cosmological constant. I don't need it anymore. Cross it out. The correct equation is the one I wrote down in the first place.

CLASSICAL VERSUS QUANTUM

So, what's the problem? In classical physics, there is no problem. In writing down an equation, you can include or exclude a term as you please. Of course, then your equation may not agree with experiments. As a classical physicist, you may craft your equation to fit experiments by adding or removing terms as needed.

Imagine two classical physicists trying to figure out an equation to describe the motion of the electron. There we are, motion of electron = charge times electric field. Do you think we should also include a term describing how the spinning electron would tumble and shake in a magnetic field? Maybe the equation should be motion of electron = charge times electric field + spin times magnetic field. Beats me, how would I know?

To decide, they would have to experiment with an electron in a magnetic field. That's more or less how classical physics proceeds.

So Einstein, the consummate classical physicist, felt perfectly free to leave the cosmological constant out. Nobody could stop him. Without the cosmological term, his original equation agrees well with observations, and that's what counts in classical physics.

Of course, if you are a classical physicist, you can always include a tiny cosmological constant, tiny enough so it has essentially no effect. Nobody can stop you, either. Astronomical observations can only tell us that the cosmological constant must be smaller than a certain value.

The difficulty comes with quantum physics. Because of quantum fluctuations, you are no longer allowed to exclude terms as you please. You must have a reason why a given term should not be there. Roughly, the reason is that quantum physics is probabilistic. Physicists can only determine the probabilities of various processes occurring. Any process not explicitly forbidden by a basic principle will occur, even though the probability of the process actually occurring may be very small.

For example, suppose that the two classical physicists did not include a term describing how a spinning electron would tumble in a magnetic field. Since no known principle forbids an electron tumbling in a magnetic field, there is some probability that the electron will do so. Thus the equation has to include a term describing that particular motion.

In quantum physics, if you neglect to include a term not specifically forbidden by any principle, quantum fluctuations would force that term on you. Indeed, the equation describing the motion of the electron was first written down without an "anomalous magnetic moment term." Later, physicists realized that quantum fluctuations forced them to include that term. Experimenters then confirmed that the term was indeed necessary to describe the electron's behavior in a magnetic field. Our confidence in the reality of quantum fluctuations has been built up through numerous episodes of this kind.

In Theodore White's *The Once and Future King,* the boy Arthur dreams of visiting a kingdom governed on the principle that whatever is not forbidden is required. The story inspired the eminent physicist Murray Gell-Mann to quip that in quantum physics what is not taboo is a commandment.

This represents a profound difference between classical physics and quantum physics.

LIKE AN EVIL GENIE

Let's recap. Einstein started out without the cosmological constant term. Then he put it in. Then again, in the face of observational evidence, he took it back out. "Worst mistake I ever made," he muttered.

Einstein could have his regrets, but the cosmological constant, once let out of the bottle like an evil genie, could not be so readily dismissed. Hey, Albert, we live in a quantum world, whether you like it or not. Once you showed us that this extra term *could* be present, you can't get rid of it. The cosmological constant term, unless we can think of some reason to forbid it, is forced on us.

You may think, gosh, all this fuss, why can't these physicists just invent some principle that forbids the cosmological constant term? The difficulty is that any principle they can think of forbids some other term. For instance, one principle that would forbid the cosmological constant also forbids the very term that describes gravity. The trick is to formulate a principle that forbids only the cosmological constant term, and nobody yet knows how to do that.

NATURAL EXPECTATIONS

Quantum physics requires that any term not specifically forbidden must be present, but it does not say what the strength of the term must be. However, physicists have a natural expectation of what the strength of a physical quantity ought to be simply by comparing what the strengths of similar physical quantities are known to be.

Without knowing any better, physicists expect similar physical quantities to have roughly the same value. For instance, all solids have roughly the same density—that is, they all weigh about the same per unit volume. For an example at a more fundamental level, the masses of particles that interact via the strong force are all roughly the same.

There is nothing particularly profound to this "principle" of natural expectation. We apply it in all sorts of contexts. Knowing that a hammer costs $9.95 and a screwdriver $7.95, you would quite naturally expect that a wrench would cost about $10, maybe $20 tops. As a more sophisticated example, consider that, as every investor knows, at any given time the price-to-earnings ratios of stocks are all roughly the same, even though the prices and earnings of stocks cover a wide range. This example points out that some sophistication is needed to determine what is meant by "similar physical quantities." Thus, in an example given above, I said "all strongly interacting particles" rather than "all particles."

Obviously, the importance of having "natural expectations" is so as to be able to spot discrepancies. Having learned the average price-to-earnings ratio, the investor can now scan the financial pages for stocks with abnormally high or abnormally low price-to-earnings ratios. A large deviation demands an explanation.

Similarly, physicists feel uneasy whenever two presumably similar quantities differ considerably in value. For instance, physicists were once puzzled by why the proton is about ten times more massive than another strongly interacting particle called the pi-meson. The eventual explanation provided one of the keys that unlocked the secrets of the strong interactions.

THE BIGGEST DISCREPANCY EVER

Having discussed what physicists mean by natural expectations, let us now return to gravity. Quantum fluctuations require a cosmological constant term to be present. But how large a cosmological constant?

Knowing the value of Newton's constant G —that is, the strength of the gravitational force—physicists have a natural expectation of how large the cosmological constant ought to be.

Now hold on to your hat! The naturally expected value for the cosmological constant comes out to be about 10^{123} times the maximum value that the astronomers tell us the cosmological constant can possibly have. This discrepancy of a factor of 10^{123} between observation and theoretical expectation has been called the biggest error in the history of physics.

Physicists are outraged. You would be similarly outraged if you found that a wrench costs 10^{123} more times than you expected. Even in the military it shouldn't cost that much!

This number, 10^{123}, is absurd. For comparison, the total number of photons in the visible universe is only (!) about 10^{88}. When invoking the principle of natural expectations, physicists are willing to accept a considerable amount of slack in what they mean by "roughly the same." Had the cosmological constant come out to be a hundred times larger than it actually is, physicists might be willing just to say to each other, Well, what is a factor of a hundred between friends?

What would the universe be like if the cosmological constant has the value physicists expect rather than the value set by Whoever designed the universe? The universe would go berserk, that is what! Rather than a stately evolution, the universe would zip through its history in no time at all. A closed universe would expand and contract in something like 10^{-44} second.

Physicists are terribly embarrassed by this huge failure of their much-trumpeted ability to anticipate Nature. The discrepancy is so large that no amount of squirming could get them off the hook. There is no plausible deniability here! Well, a physicist might say, perhaps it is not right to base our expectation for the cosmological constant on Newton's constant, since we know for sure that we don't understand quantum gravity and the physics at the Planck energy. All right, suppose by fiat we ignore the quantum fluctuations generated by gravity. But quantum fluctuations generated by the other three forces still would produce a cosmological constant. For instance, physicists expect the cosmological constant produced by the strong force to have a value 10^{51} times larger than the maximum value allowed by the astronomers. (This value is obtained by considering the value of a certain quantity characteristic of the strong force.) Well, a discrepancy of 10^{51} is quite an improvement over one of 10^{123}, but it is still ridiculously large.

Now, *that*'s a paradox.

A FAILURE OF PHYSICS

But what about the superstring theory? you wonder. Don't superstringers claim to have a new understanding of gravity?

When the superstring theory first came out, physicists waited with bated breath to see how the cosmological constant would come out. There was an initial flurry of excitement: When the theory was first written down, it looked as if the cosmological constant might come out to zero. But unfortunately, that proved not to be the case.

The genie is still out of the bottle. Rats!

The failure of the superstring theory to solve the cosmological constant paradox is one of the reasons why the infidels do not accept its claim as the ultimate theory, at least for the moment. Meanwhile, the faithful are hoping for a fourth miracle.

None of the other ideas about gravity has anything illuminating to say about the paradox, either.

Some physicists hope that unknown dynamics may push the cosmological constant toward zero and keep it there. Consider, as a rough analogy, a marble rolling in a ditch. The marble will end up exactly at the bottom and stay there, because of friction. The marble being at the bottom is the analog of the cosmological constant being equal to zero, and the combination of the marble's weight and friction is the analog of the unknown dynamics.

The notion that dynamics may drive the cosmological constant toward zero and keep it there is intriguing, but thus far it remains a hope. Nobody knows what the dynamical mechanism might be. There is a rather vague suggestion that the effect of a humongous cosmological constant on the universe may be so violent that the universe may strike back. In other words, space and time curved and stretched too violently may react by producing effects that may cancel the effect of the cosmological constant. Well, if you think this is kind of vague, it is.

Other ideas—some of them really wild—are bubbling in the theoretical physics community. But they are all too uncertain to be discussed here.

There is nowhere to hide from the cosmological constant paradox. We can't hide behind our ignorance of the physics beyond the Planck energy. As explained above, even the physics we claim to understand, such as the strong force, already generates a humongous cosmological constant.

There is something seriously missing from our understanding of gravity.

WE WANT TO LAUGH, TOO

Man thinks, God laughs.

—JEWISH PROVERB

And thus we come to the end of our dance with gravity, perplexed and puzzled.

We thrilled to the freedom of free-fall. Yet we still do not understand completely why we fall.

Physics began with gravity, and ever since, physicists have struggled to understand its innermost secrets. Dropping cannonballs from the Leaning Tower of Pisa, Galileo discovered that gravity acts on all things equally, a fact with implications far beyond his wildest imaginings. An overripe apple hit the daydreaming Newton on his head, and he brought heaven down to earth. By quantifying gravity as a universal force, Newton showed that celestial mechanics was no different from terrestrial mechanics. While the apple fell, the moon also was falling. The patent clerk stumbled upon his happy thought. Did the daydreaming Einstein actually tip over in his chair with a resounding crash? Did he pat the lump on his head and understand what gravity was about? With an astonishing insight, the patent clerk saw the connection between Galileo's cannonballs and the fabric of space and time. Searching for a sensible quantum theory of gravity, physicists came upon a vision of the universe made of itsy-bitsy pieces of string. Through gravity, physicists may yet get to know all else.

Physics began with gravity, and it may end with gravity. The attempt to understand gravity has brought physicists face to face with the biggest paradox of all time. The cosmological constant paradox is big. A lawyer would argue that without an understanding of the paradox, the rest of physics does not make any sense. Here, you say that you've got this beautiful theory that explains the physical world, but in your theory you also expect one of the terms to be 10^{123} times larger than the observed value!!! How then can I trust your explanation? Whom are you trying to kid?

If you ever see a fundamental physicist banging his or her head against the wall in frustration, ask him or her if it is because of the cosmological constant paradox.

Much of what we do not understand of gravity can be traced to its frightful clash with the quantum. Were the world not quantum, gravity would have been happy. Einstein's theory would be wonderful and consistent. But no, quantum fluctuation has to wreak its havoc, and it does because gravity is propelled by mass and energy.

The mystery is the pitiful weakness of gravity. Sisyphus in his labor would not have thought so, but compared to the other forces, gravity appeared to be a teeny afterthought in the design of the universe. The curious twist is that we know about this absurdly feeble force by the happenstance that huge numbers of particles have come together to form macroscopic objects. But who brought these macroscopic structures out of an otherwise inchoate universe? Why, none other than gravity!

The flip side of gravity's weakness is its stupendous characteristic energy—we have to go up a long way on the energy scale before gravity becomes comparable to the other forces in strength. As I said before, gravity is a mysterious interloper from an energy realm almost inconceivably remote from our low-energy existence. Can we ever understand the physics at energies of 10^{19} GeV and beyond? Can we see the universe at its creation, a beginning now shrouded in the epoch of dark ignorance?

We don't even know whether the answer is contained within

existing physics or whether drastically new physics is needed. I am hoping that the resolution of the paradox will lead to a new revolution comparable to that of the quantum. Is the answer to be found within the dark secrets of gravity? Or does it lie in the jittering quantum? The quantum principle was uncovered in the atom with its characteristic energy scale of a few electron volts and has been verified up to the energy scale of contemporary particle physics of 1,000 GeV or so. Would it continue to hold at 10^{19} GeV?

Physics began with gravity, and it may end with gravity.

"Man thinks, God laughs," so the proverb says. We are thinking, but we want to laugh, too. Are we on the threshold of knowing His thoughts, to use Einstein's phrase, or are our theories still laughably childish? Will the day ever come when, with our thinking done, we can join the laughing God and have a good chuckle over how cleverly the universe is designed?

Note: In the spring of 1988, some physicists suggested a possible resolution of the cosmological constant paradox. They think that the cosmological constant may be effectively driven to zero by quantum fluctuations in the fabric of space and time. While this proposal has generated a great deal of excitement, it remains to be seen whether it is viable.

Notes

PROLOGUE

Page

xv.   For biographical material on Newton, I have used *Never at Rest: A Biography of Isaac Newton*, by Richard S. Westfall (Cambridge, Eng.: Cambridge University Press, 1980). For a fascinating psychoanalytic study, see *A Portrait of Isaac Newton*, by Frank E. Manuel (Cambridge, Mass.: Harvard University Press, 1968).

xvi.   Manuel, op. cit., speaks of Newton's fixation on his mother, to whom he was devoted. Newton was secretly an anti-Trinitarian and had religious delusions. According to Manuel, Newton even believed that he was the only son of God and could not endure the rivalry of Christ. In his alchemical writings, Newton had used the name "Ieova sanctus unus," an anagram of his Latin name Isaacus Neuutonus. The quest for father took on a number of other guises as well.

xvi.   Newton left quite a bit of material for psychoanalysts. As part of his schoolwork, Newton was to write a list of words under each letter of the alphabet. Often he would deviate from the official text and freely associate. Under *M* he wrote Marriage, Mother, Marquesse, and Manslayer; under *W*, Wife, Wedlock, Wooer, Widdow, Widdower, and Whoore; under *F*, Father, Fornicator, Flatterer, Frenchman, and Florentine. (The last two *F* words were then used to describe men of loose morals.) His hatred of his half brother Benjamin was manifest in the list of words under *B*: Brother, Bastard, Barron, Blasphemer, Brawler, Babler, Babylonian, Bishop, Brittaine, Bedlam, Beggar, Brownist, and Benjamite. Incidentally, Newton's mother died from a fever she contracted while taking care of the sick Benjamin.

xviii.   Actually, in his notes of 1666, Newton did not even clearly speak of gravity as a force. Westfall demonstrates convincingly that the popular portrayal of Newton's discovery of gravity as an almost divine inspiration is by and large a myth.

xix. Galileo's experiment was actually first done in the sixth century A.D. by Philoponus of Alexandria and later, starting in the fourteenth century, by Jean Buridan, Leonardo da Vinci, and Simon Stevin. See Edward Harrison, *Masks of the Universe* (New York: Macmillan Publishing Company, 1985), page 85.

xxiii. Note that the acceleration of a planet, being equal to the gravitational force exerted by the sun on the planet divided by the planet's mass, is equal to GM/R^2. It does not depend on the planet's mass. Here M denotes the sun's mass and R the radius of the orbit. Evidently the combination GM can be determined from the motion of the planets. If we can measure the value G in the laboratory (by carefully measuring the tiny attractive forces between two metal balls of predetermined masses), then we can obtain the value of M and thus in effect have weighed the sun. Clever, eh? In Newton's time, laboratory techniques were not refined enough to measure G directly. Nevertheless, it was possible to estimate the value of G. Apply the reasoning above to the acceleration of falling objects and we see that the known value of thirty-two feet per second per second determines the combination Gm, where m denotes the earth's mass. Knowing the size of the earth and the density of rock, we can estimate the earth's mass and hence obtain an approximate value for G.

CHAPTER 1

Page

3. E. Rogers in *Einstein: A Centenary Volume,* edited by A. P. French (Cambridge, Mass.: Harvard University Press, 1979). Figure 1.1 is based on a drawing by Rogers.

4. I have drawn upon two Einstein biographies: Ronald W. Clark, *Einstein: The Life and Times* (New York: World Publishing Company, 1971), and Abraham Pais, *Subtle Is the Lord . . .* (New York: Oxford University Press, 1982).

11. I. B. Cohen's account of his visit to Einstein appeared in *Scientific American,* vol. 193, pp. 69–73 (1955).

11. To study various processes in the absence of gravity, scientists now routinely drop their experiments from high towers or conduct them in diving airplanes. Einstein's happy thought has now become commonplace. See J. Walker, *Scientific American,* vol. 254, p. 114 (1986).

12.   The candle flame story may be found in the reminiscences of Banesh Hoffman in *Albert Einstein: Historical and Cultural Perspectives,* edited by Gerald Holton and Yehuda Elkana (Princeton, N.J.: Princeton University Press, 1982).

12.   Here is another one of those puzzles Einstein would have liked. A child sitting on a train is holding a helium balloon. As the train suddenly accelerates forward, which way does the balloon move?

12.   Some of the notable figures who repeated "Galileo's experiment" include Baron Eötvös in the nineteenth century, and Robert Dicke and Vladimir Braginsky in recent years. Physicists are trying constantly to improve the accuracy with which this fundamental experiment is performed. Recently, Eric Adelberger and others have been working hard on improving this experiment.

13.   Strictly speaking, even if you are deep in space, far away from any massive object, you will feel a small but not insignificant gravitational force. Although each massive object in the universe will exert only an infinitesimal force on you, there are so many objects in the universe that the force adds up to a small but not zero value.

16.   The passage is taken from Aaron T. Beck and Gary Emery with Ruth L. Greenberg, *Anxiety Disorders and Phobias* (New York: Basic Books, 1985), p. 210. "Some acrophobics fear they may have a perverse uncontrollable urge to jump and may even feel as though some *external force is drawing them to the edge* of the high place. An acrophobic may have visual fantasies of falling and even experience bodily sensations of falling despite being firmly situated on solid ground. The sensations of falling . . . are forms of *somatic imaging . . .*" (their italics). These authors go on immediately from a discussion of acrophobia to elevator phobia. The elevator phobic fears that "the cables will break and the elevator will crash." It may be relevant to note that the rigged living room in our discussion was actually an elevator in Einstein's original discussion. Was Einstein an acrophobic and an elevator phobic? Did the theory of gravity originate from fear?

CHAPTER 2

Page

18.   For the interested reader, an extraordinarily clear (but highly technical) discussion of the equivalence principle in its various forms may be found in Steven Weinberg's *Gravitation and Cosmology* (New York: John Wiley & Sons, 1972), pp. 91–93.

18. Of course, we have confidence in the equivalence principle ultimately because of the many experimental tests physicists have subjected it to.
They are continuing their tests with ever more refined techniques. See the
seventh note to Chapter 1.

22. Of course, we can make an object appear to "fall" upward by using
another force, such as the electric force, to overwhelm gravity. Thus, if the floor
is electrically charged, it can push an electrically charged object upward.

22. In reporting a unsubstantiated speculation on the possible existence of a fifth force, some articles in the popular press stated that the equivalence
principle may fail. This is a rather misleading way of putting it. The equivalence
principle is a statement about gravity. See the preceding note. (As of this writing,
the evidence, particularly from the experiment of Eric Adelberger and his collaborators, appears to be against the existence of a fifth force.)

26. I read the princess-in-the-tower analogy so long ago, in high
school, that I can no longer remember where it appeared. I believe that it is due
to Gamow.

29. A fuller discussion of symmetry, the impact of the equivalence
principle, and of Einstein's theory of gravity may be found in my book *Fearful
Symmetry* (New York: Macmillan Publishing Company, 1986).

CHAPTER 3

Page

39. Hertz felt that the wave he detected could not possibly be of any
practical use.

40. The funding requested for the gravity-wave detectors amounts to
twenty-five cents per person.

42. For a more extensive introduction to quantum physics than can
possibly be given here, see Heinz R. Pagels, *The Cosmic Code* (New York: Simon
& Schuster, 1982) and my *Fearful Symmetry*.

43. I trust that Aristotelian scholars will not be too upset by my parody
of Aristotle's physics.

CHAPTER 4

Page

50. The building of the Mount Wilson Observatory is told in a marvelously vivid way in David O. Woodbury, *The Glass Giant of Palomar* (New York: Dodd, Mead & Company, 1939, 1960). The reminiscences of Jerry Dowd may be found here.

50. Another interesting book on telescopes is the one by R. Lerner, *Astronomy Through the Telescopes* (New York: Van Nostrand Reinhold, 1981). See also Henry C. King, *The History of the Telescope* (London: C. Griffins & Co., 1955).

50. George Willis Ritchey (1864–1945) studied at the University of Cincinnati. When Hale hired him, he was an instructor in the Chicago Manual School. Woodbury implies that Ritchey left Mount Wilson half insane. According to the *Dictionary of Scientific Biography* (to be abbreviated as *DSB*), edited by C. C. Gillispie (New York: Charles Scribner's Sons, 1957), Ritchey was dismissed from Mount Wilson after a bitter controversy, partly because he was alleged to have exceeded his authority and partly because of what the *DSB* delicately referred to as "a health problem." After his departure from Mount Wilson, Ritchey continued to function as a preeminent telescope maker. In particular, the French decorated him for helping them build a telescope.

52. For biographical data on Milton La Salle Humason, see *Biographical Encyclopedia of Scientists,* edited by J. Daintith, S. Mitchell, and E. Tootill (New York: Facts On File, 1981) and *Asimov's Biographical Encyclopedia of Science and Technology,* by Isaac Asimov (Garden City, N.Y.: Doubleday & Company, 1982). Humason died on June 18, 1972.

53. In his article in the *DSB*, G. J. Whitrow suggested that Hubble's legal training helped him in sifting through the mountain of astronomical data.

54. Figure 4.1 is taken from Kenneth Glyn Jones, *The Search for the Nebulae* (Halfpenny Furze, Mill Lane, Chalfont St. Giles, Bucks, England: Alpha Academic, a division of Science History Publications; U. S. distributor: Academic Publications, 156 Fifth Avenue, New York, N.Y. 10010).

56. For a more detailed history of the discovery of the expansion of the universe, with reference to the original work of Slipher, Wirtz, Lundmark, Hubble, and Humason, see Steven Weinberg, *Gravitation and Cosmology* (New York: John Wiley & Sons, 1972) and *The First Three Minutes* (New York: Basic Books, 1977). Weinberg expresses doubt that Hubble could have extracted his 1929 conclusion from the then-available data.

56. To apply his principle, Johannes Christian Doppler (1803–53) arbitrarily assumed that all stars were intrinsically white and emitted only light in the visible part of the spectrum. Buys Ballot criticized Doppler on making this assumption. The Doppler effect apparently was also discovered independently in 1848 by J. S. Russell and by H. Fizeau. Incidentally, Christoph Hendrick Diedeck Buys-Ballot was an eminent meteorologist (like Friedmann, see Chapter 6). See the *DSB*.

56. As we saw in Chapter 3, time is warped in a gravitational field. Thus, the frequency of light passing through a gravitational field will be shifted. Light emitted from astronomical objects is shifted toward the red. This gravitational redshift must be subtracted from the redshift due to motion.

64. Edward Harrison (in *Physics Today*, February 1986) makes the intriguing suggestion that Newton did worry about the dynamics of the universe but suppressed his calculations for fear of incurring ecclesiastical wrath.

67. An important distinction between Newtonian and Einsteinian cosmology is that in the former the galaxies are racing apart through a static, unchanging background of space, while in the latter space itself is expanding. The raisin bread analogy is closer to the Einsteinian model: Space, namely the bread between the raisins, is itself a dynamic entity.

69. For biographical material on Lemaître, see A. Deprit, "Monsignor Georges Lemaître" in *The Big Bang and Georges Lemaître* (Dordrecht, Holland: D. Reidel Publishing Company, 1984). I would like to thank Deprit for an interesting correspondence.

69. From patent clerk to physicist, from country lawyer to astronomer, from mule driver to janitor to astronomer, from meteorologist to cosmologist, from mining engineer to mathematician, from abbé to cosmologist—given all these examples, are you thinking of a change of career?

CHAPTER 5

Page

74. Edward Harrison, *Cosmology: The Science of the Universe* (New York: Cambridge University Press, 1981). In this book, Harrison went a long way toward clearing up the confusion surrounding the night sky paradox. The resolution of the paradox does not require an expanding universe, as is often erroneously stated. This is Kelvin's point.

74. D. G. Clayton, *The Dark Night Sky* (New York: Times Books, 1975).

77. For a detailed account of how our conception of the universe evolved, see Edward Harrison, *Cosmology: The Science of the Universe.*

77. Harrison has traced the night sky paradox past de Cheseaux back to Thomas Digges in 1576, Johannes Kepler in 1610, Otto von Guericke in 1672, and Edmund Halley in 1721.

77. Edmund Halley of comet fame remarked in 1720: "If the number of the Fixt Stars were more than finite, the whole superficies of their apparent Sphere would be luminous." It is not clear to me whether he simply envisaged all the stars affixed to the celestial sphere at a constant distance from us.

77. Johannes Kepler railed against the notion of an infinite universe, perhaps partly because of his realization of the night sky paradox.

77. It has been established that Olbers actually owned a copy of de Cheseaux's book, but it is doubtful if he ever made it to the appendix. See F. P. Dickson, *The Bowl of Night* (New York: Philips Technical Library, Cleaver-Hume Press [Macmillan & Co.], 1968).

79. Steven Weinberg, *The First Three Minutes,* p. 29.

80. The material about Lemaître and his unease at Pope Pius XII's pronouncements may be found in A. Deprit, *The Big Bang and Georges Lemaître.*

80. The reference to Jeans comes from the following passage from Sir James Jeans, *The Mysterious Universe* (New York: Macmillan & Company, 1932), p. 142: "Or again, it may be that our consciousness should be compared to the feeling in the finger of the painter as he guides the brush forward over the still unfinished picture. If so, the impression of influencing the parts of the picture yet to come is something more than a pure illusion."

81. D. W. Sciama once remarked: "Olbers could have predicted the expansion of the universe, and even made a rough estimate of Hubble's constant, a hundred years ahead of the observers. His failure to do so is one of the greatest missed opportunities in the whole history of science" in *The Unity of the Universe* (London: Faber & Faber, 1959). This rather provocative remark was examined critically by Peter T. Landsberg and David A. Evans in "What Olbers Might Have Said" in *The Emerging Universe,* edited by William C. Saslaw and Kenneth C. Jacobs (Charlottesville: University Press of Virginia, 1972). They concluded that it was rather unlikely that Olbers could have anticipated the expansion of the universe but that he might have realized that a finite age for the universe would have resolved the night sky paradox. Among other things, they pointed out that the Doppler effect was not predicted until two years after Olbers's death.

81. Two good books on cosmology are J. Silk, *The Big Bang* (San Francisco: W. H. Freeman, 1980), and H. Pagels, *Perfect Symmetry* (New York: Simon & Schuster, 1985).

CHAPTER 6

Page

85. Aleksandr Friedmann (1888–1925) was a meteorologist. His 1922 paper on the expanding universe appears to be a digression in his career. In 1914 he volunteered to be an aerial observer in the Russian Air Force.

86. We walk around supporting on our head a column of air several miles high, reaching all the way to the edge of the earth's atmosphere. This air column weighs about fifteen pounds for every square inch on your head. This so-called atmospheric pressure decreases as we climb up simply because the air column gets shorter and shorter. Imagine filling a perfectly elastic bag with air. If we carry this bag up a mountain, the bag will expand simply because there is less atmospheric pressure bearing down on it.

86. In discussing the expansion of the gas, I assumed that the gas does not exchange energy with its environment. Technically, this is known as an adiabatic expansion.

90. Much of the material on Gamow's life was taken from George Gamow, *My World Line* (New York: Viking Press, 1970). S. Ulam once described Gamow as "Perhaps the last example of amateurism in scientific work on a grand scale."

90. Benjamin Franklin defined man as a tool-making animal. Perhaps he was a dialectical materialist.

90. In *My World Line,* Gamow stated that Friedmann died of pneumonia. However, according to the *DSB,* Friedmann died of typhoid fever. All things being equal, one might be tempted to trust Gamow, since he was right there. However, Gamow was notoriously cavalier with details, and he might have said pneumonia just to make it a better story. I have followed Gamow here.

91. There is a curious irony in the fact that Gamow's work in quantum physics concerns the escape of the so-called alpha particle from the confinement of certain atomic nuclei.

93. During the epoch before nucleosynthesis, the universe also contains, besides protons, neutrons, electrons, and photons, some exotic particles such as neutrinos, which need not concern us here. See Chapter 10, however.

95.   The poem by Dickinson exists in more than one version.

95.   Another reason why the early universe is easy to describe is because the physics involved is essentially classical rather than quantum. In following the encounters between the different particles, we need only note the various reaction rates. Thus, in many ways the early universe is easier to understand than a piece of metal, a point that surprises many nonphysicists. (The properties of metals can be understood only in terms of quantum physics.)

96.   It so happens that there is no stable nucleus with five nucleons— that is, five protons and neutrons. Thus, after the helium nucleus (which consists of four nucleons) was formed, nucleosynthesis was faced with the so-called A-5 bottleneck. Were there no bottleneck, the helium nucleus in encountering a proton or a neutron could have formed a nucleus with five nucleons, which in turn could form a nucleus with six nucleons, and so on. Primeval nucleosynthesis would have progressed rapidly. To get past the bottleneck, the helium nucleus had to find another helium nucleus to form a nucleus with eight nucleons. The early universe contained too few helium nuclei for such encounters to occur frequently. Thus the buildup of "higher" nuclei had to wait till the birth of stars.

100.   The discovery of the cosmic background radiation represents one of the most fascinating chapters in the history of physics. Why was the discovery not made earlier? Why was Gamow's contribution forgotten? These and other questions are explored by Weinberg in *The First Three Minutes* and by J. Bernstein in *Three Degrees Above Zero* (New York: Charles Scribner's Sons, 1984). Of special significance and interest is a first-person account by Ralph A. Alpher and Robert Hermann, "Reflections of Early Work on 'Big Bang' Cosmology" (*Physics Today*, August 1988, page 24).

CHAPTER 7

Page

106.   When we spoke of putting some protons and neutrons into a box, we were speaking of an imaginary box, of course.

107.   We thank Maurice Goldhaber for telling us about his joke.

107.   Some readers may be puzzled by the question of why the neutrons inside atomic nuclei do not disintegrate. Some nuclei—the radioactive ones—do disintegrate, but many are stable. The nuclear force that binds neutrons inside the nuclei also prevents them from disintegrating.

107. By studying ancient rocks of known geological age, one can set a far better lower limit on the proton's lifetime than by the mere observation that we do not glow.

108. Strictly speaking, since we do not know whether the universe is finite or infinite, we should not speak of the number of baryons in the universe. Rather, physicists speak of the number density of baryons—that is, the number of baryons per unit volume in the universe. If the universe is infinite, then the total number of baryons would be infinite, of course. When physicists do talk about the total number of baryons, they mean the total number within the visible universe.

112. Up till the mid-1970s, particle physicists typically knew little about cosmology. I believe that Steven Weinberg's popular 1977 book *The First Three Minutes* had considerable influence in introducing particle physicists to cosmology.

120. We focused on the genesis of baryons. You may well wonder where electrons came from. In the grand unified theory, electrons and baryons are intimately linked together, as mentioned in the text. The same processes that generated baryons also generated electrons and particles related to electrons.

120. Clearly, if the genesis of matter were to work according to the scenario outlined in the text, the universe had to expand significantly faster than the rate at which processes violating baryon number proceeded. Thus, baryons were generated only when the universe was expanding rapidly.

120. Long before grand unification, the great Soviet physicist and humanitarian Andrei Sakharov speculated that the origin of matter in the universe may be understood if the number of baryons is not strictly conserved. Few in the West were aware of Sakharov's work. After a grand unification was proposed, a number of physicists, including M. Yoshimura, S. Dimopolous, L. Susskind, D. Toussaint, S. Tremain, F. Wilczek, S. Weinberg, and I reinvented the scenario for the genesis of matter and placed it in the framework of grand unification.

122. George Gamow made the crack about St. Augustine in his book *The Creation of the Universe* (New York: The Viking Press, 1952, 1961).

122. St. Augustine, *Confessions,* translated by R. S. Pine-Coffin (Penguin Books Ltd., 1961, Book XI).

CHAPTER 8

Page

124.   A characteristic property of gravity, that the force is attractive between any two bits of matter, plays a crucial role here. In contrast, the electromagnetic force is attractive between opposite charges and repulsive between like charges.

CHAPTER 9

Page

134.   A question: "Is it possible to have two whorls on your head?"

140.   Vortices occur only in certain types of superconducting metals. The Russian physicist A. A. Abrikosov was the first to recognize that vortices can exist in superconductors.

140.   The modern theory of superconductivity was worked out by J. Bardeen, L. Cooper, and J. R. Schrieffer. They were awarded the Nobel Prize for Physics in 1972 for this work.

140.   This chapter was written some time before the current excitement over materials that superconduct at relatively high temperatures. The mechanism for superconductivity at high temperature may well be different from what was described in the text. The discussion in the text, intended to illustrate a point about topology, is not affected in any way.

142.   The groundwork for the revolution in fundamental physics mentioned in the text was built up slowly over the years. I have to brush over a number of technical details. More precisely, the arrows represent the so-called Higgs fields pointing in an abstract internal space. See my book *Fearful Symmetry* for a fuller explanation.

143.   The work of 't Hooft and Polyakov inspired the French physicist Bernard Julia and me to make a small contribution to this subject. We showed how to construct a hedgehog that emits an electric as well as a magnetic field, an object known as a dyon.

145.   Eventually, as the expansion of the universe slowed down, light brought different regions of the universe into touch. In physicists' language, we humans are in touch with a distant galaxy by the very fact that we see the distant galaxy. But even now, we may be out of touch with regions so remote that light

emitted there, zinging across the universe ever since the big bang, has yet to reach us.

145.   Cultural historians love to debate to what extent various ancient civilizations were out of touch. In a fascinating recent book, the eminent sinologists Joseph Needham and Gwei-djen Lu demonstrated convincingly that various cultural and technological artifacts appeared in Meso-American civilizations approximately two centuries after they appeared in East Asian civilizations. With the earth not expanding, the Kuroshio Current could have carried ancient travelers across the Pacific. But to a large extent, civilizations in the ancient world developed independently.

148.   A. Linde, P. Steinhardt, A. Albrecht, and K. Sato are among the theorists who contributed to the development of the inflationary scenario.

149.   Stretching out the universe by an enormous factor, inflation effectively homogenized the universe. Thus, to a large extent, the evolution of the universe after the inflationary epoch does not depend on the precise condition the universe was in before inflation. Some physicists find it appealing that in order to explain the universe as we know it now, we do not have to know the so-called initial conditions of the universe. Others, on the other hand, find it disturbing to think of an amnesiac universe which has totally forgotten what it was like before inflation.

151.   *Cosmic string* is not to be confused with the *superstring*, to be discussed in Part 4 of this book.

151.   The astute reader may have realized that the arrows in a hedgehog are three-dimensional while the arrows in a vortex line or cosmic string are two-dimensional. Actually, these arrows exist in an internal space, not the space we live in.

153.   Black holes also could work as gravitational lenses, but a cosmic string would be expected to produce a line of double images.

CHAPTER 10

Page

162.   M. Aymé, Le Passe-Muraille (Paris: Librairie Gallimard, 1943).

171.   You may wonder how the neutrino could ever be detected. The answer is that you pile an enormous amount of matter (a junked battleship is one favorite) in front of a neutrino beam produced by an accelerator and wait pa-

tiently. To produce a neutrino beam, experimenters first produce a beam of particles called *pi mesons,* which are known to decay into the neutrino plus another particle. Actually, there are three different kinds of neutrinos.

171.   Strictly speaking, it is not ruled out that the neutrino could have a teeny magnetic moment, for instance, so that it would have an exceedingly small interaction with an electromagnetic field.

172.   At present, the experimental situation on the neutrino's mass is rather confused. In the past few years, a Russian group and a Canadian group have separately claimed to have measured a neutrino mass. However, subsequent experiments have failed to confirm these claims. At present, experiments indicate that if the neutrino has a mass at all, its mass has to be less than the electron's mass by a factor of twenty-five thousand.

174.   Frank Wilczek, together with his collaborators Blas Cabreras, John Moody, and Larry Krauss, and, independently, Pierre Sikkivie have studied how available technology could be pushed to detect the axion and other dark matter particles. A number of other physicists have worked in this area.

CHAPTER 11

Page

179.   I spoke of two extreme possibilities for the dark matter particle: The wimp may have a tiny mass and each volume of space contains lots and lots of this particle, or the wimp may be very massive and each volume of space contains only a few wimps. These two types of dark matter particles are known as hot and cold, respectively. The terminology comes from how fast the wimp was moving at the time when structures first emerged.

179.   In the classification of the preceding note, the axion has the properties of a cold dark matter particle.

184.   R. Dicke and J. Peebles argued for a long time that there are only three aesthetically attractive values for Ω: zero, one, and infinity. Of these, only $\Omega = 1$ gives rise to an interesting universe.

184.   A more serious argument is based on how the value of Ω changes as the universe evolves. Both the actual mass density and the critical mass density of the universe change as governed by Einstein's theory of gravity. It turns out that if $\Omega = 1$ exactly, then Ω will stay equal to 1. In contrast, if Ω is larger than 1, Ω will increase, and if Ω is less than 1, Ω will decrease. In other words, only

the value $\Omega = 1$ is stable. What does all this imply? Suppose observation shows that Ω is actually 1. Fine. That would mean that Ω has always been 1 since the beginning. In contrast, suppose the observation shows that Ω is 0.2, say. Well, since we just said that Ω decreases if it is less than 1, Ω must have been decreasing steadily from a value slightly less than 1. Going backwards in time, we see that in the early universe Ω would have to be ever closer to 1. The thought that He would have started out the universe with Ω exceedingly close to 1, but not quite equal to 1, disturbs a lot of physicists.

186.   When we look into a night sky and see moving lights, we assume that light traces mass and that an airplane is attached to each spot of light.

187.   With rapidly advancing technology and experimental sophistication in measuring the cosmic microwave background, many of the questions discussed in this and the preceding chapter should be answerable in the near future.

CHAPTER 12

Page

201.   Calculation of the effect of quantum fluctuations on forces between particles is a subject known as renormalization theory.

201.   Some textbooks, particularly older ones, and hence some popular books, give the misleading impression that the quantum fluctuations to the electric force also give an infinite effect.

203.   Corresponding to the Planck energy is a length scale of about 10^{-33} centimeters, known as the Planck length, naturally. The string is Planck length in size.

209.   It is reasonable to suppose that photons are generated as the universe passes through death to rebirth. In any process in which particles crash together, photons are invariably emitted. In contrast, the number of protons (baryons, strictly speaking) is conserved. (Physicists believed in baryon number conservation when the argument was advanced.) Thus, in every cycle the number of photons per proton in the universe is larger than in the preceding cycle. How can that be? the argument goes. Since the universe has no beginning and has gone through an infinite number of cycles, the number of photons per proton now ought to be huge, if not downright infinite. Hence, ha! The universe can't possibly be cyclic.

Let us not debate whether the argument is correct. It was only meant to be suggestive. In any case, the argument has evaporated now that most physicists believe in the prediction of the grand unified theory that the number of protons is not conserved. As I explained in Chapter 7, the number of photons per proton is fixed by the laws of physics. It is, of course, possible, given that we know nothing about what went on in the era of dark ignorance, that the laws of physics themselves could be changed as the universe passes through death to rebirth. But there is no longer any argument suggesting that the number of photons per proton increases from cycle to cycle. In summary, if you gave up on the cyclic universe because you accepted the absolute conservation of baryon number and you bought the argument presented above, you can once again believe in life after death for the universe.

CHAPTER 13

Page

212. In this long footnote, I sketch the remarkable history of the superstring theory.

The emergence of the string theory offers a superb illustration of the fact that ideas in physics often (but not always) come out of left field. While some physicists were banging their heads begging gravity to marry the quantum, others in the late 1960s were advocating a radical approach to describe the behavior of particles that interact via the strong force, particles called hadrons. At that time, the research of these strong-interaction physicists appeared to be as remote from gravity as would have been possible. There is a moral here for government funding agencies who want to pigeonhole research in basic science.

These radicals wanted to give up deducing the behavior of hadrons from an underlying theory; instead, they simply tried to guess what the behavior had to be in order to agree with what was known experimentally. This rather radical approach was born out of a decade or so of frustration over the inability of quantum field theory to account for the behavior of hadrons.

Finally, the Italian physicist Gabriele Veneziano managed to produce a guess, which as far as the interaction of two hadrons is concerned agreed more or less with everything that was known. But what theory would produce Veneziano's guess as the answer? You guessed the answer to a puzzle but you can't figure out the reasoning that would lead you to the answer. Remarkably, several physicists recognized that Veneziano's rather bizarre guess emerged from a string theory.

It is one of those accidents of history that at that very moment—about 1970—quantum field theory came roaring back. In my book *Fearful Symmetry* I called the return of quantum field theory "the revenge of art," and it was a stirring story indeed. Within a few years, physicists discovered that the strong, the electromagnetic, and the weak forces could all be described by quantum field theory, and furthermore, a unified description of these three forces can be achieved by using quantum field theory.

Naturally, fewer and fewer physicists were interested in the string theory. The decline of the string theory was based on psychosocial as well as physics reasons. Everyone wanted to go with a winner, of course. Besides, since most physicists were trained in quantum field theory, the string theory looked unfamiliar and hence yucky. It was also significantly more complicated. At the same time, the string theory ran into a number of difficulties. The original version was consistent only if space and time were twenty-six–dimensional, a more refined version only in ten-dimensional space and time. Furthermore, the theory predicted a massless hadron with definite properties but that did not and does not exist.

During the 1970s, quantum field theorists (like myself) got jobs; string theorists didn't, or they got jobs that were less than what they desired. One physicist I knew agonized over whether he should mention on his job application his papers on the string theory. John Schwarz, now revered as a leading light in the development of the superstring theory, was not given a faculty position at Caltech, where he toiled for more than a decade as a research associate while people junior to him were given professorships. Nowadays it is the opposite, of course. Mention that you are a superstringer and you will be given a job. Young particle physicists who do not work on the superstring theory are regarded with suspicion. The physics community, as a sociological entity, is just as susceptible to the changing winds of fashion as society at large, naturally.

The good news is that despite the unpopularity of the string theory, a small band of determined practitioners kept the faith alive and refined the string theory into the superstring theory. With the discovery by Schwarz and Green that the superstring theory is not afflicted with the anomaly, interest revived dramatically.

216.   To be fair, I should say that Grassman numbers were already used in quantum field theory long before the advent of superstrings. However, I did want to emphasize that it is not strictly correct just to think of the bits of string wriggling in space.

220.   The Kaluza-Klein theory and the superstring theory suffer from the same difficulty: The internal space has to be maintained at a fixed size, while, as we know, the external space is expanding.

221.   The torus has an additional symmetry besides what was mentioned in the text. I should be clear, however, that the sphere is more symmetric than the torus.

224. I should also say that superstring theory can specify the types of internal space that can be used—in particular, that the internal space cannot be symmetric like a sphere comes out of the theory.

225. Superstring theory also can be described as a field theory in a mathematical two-dimensional "space" and "time." The ordinary space and time that we know are then represented as fields. Strange, eh?

226. Many physicists believe that the ultimate theory should be unique. Superstring theory fails this criterion at the moment: A number of different versions exist.

226. Those readers who want to learn more about string theory will find an article by G. Taubers ("Everything's Now Tied to Strings," *Discover,* volume 7 [November 1986], page 34) enlightening.

CHAPTER 14

Page

233. A note for the expert: In writing down Einstein's equation, I expressed the distribution of mass and energy in units used by gravitational theorists. In other words, I absorbed a factor of $8 \pi G$, where G is Newton's constant. I have also reversed the sign of the cosmological constant from its usual convention.

237. Some physicists say that the cosmological constant may be renormalized to zero. This is just a fancy way of saying that the cosmological constant can be adjusted to zero. The paradox remains concerning why it is so small.

237. During the 1920s and 1930s, some cosmologists, notably Abbé Lemaître and Sir Arthur Eddington, would include a small cosmological constant term in their equations. The resulting cosmology has some interesting features. Suppose the cosmological constant is so small that the universe starts out with the effect of matter dominating the effect of the cosmological constant. The universe expands almost as if there is no cosmological constant. Because the universe expands, the density of matter drops steadily so that at some point the cosmological constant is able to balance the effect of the matter. The universe enters a coasting period in which it stays more or less static, as envisaged by Einstein. But eventually the cosmological constant wins, and the universe enters into an explosive phase in which it expands exponentially.

237. It has also been suggested that a small cosmological constant may account for the observations that point to the existence of dark matter.

237. The financial community's use of the price-to-earnings ratio is slightly unfortunate from a mathematical point of view, since in computing the ratio, one risks dividing by zero. When the reported earnings are zero or tiny, the financial listings sometimes enter a symbol meaning "huge."

239. A long footnote for those who want to know more about the cosmological constant paradox. One way of expressing the paradox is to say that compared with other physical quantities, the cosmological constant, as measured by astronomical observation, is almost mathematically zero. Physicists know of only one way in which a physical quantity can be mathematically zero and that is because of a symmetry.

To explain this point, let me resort to a geometric analogy. Take a geometric figure such as the one in N.1a. Draw a vertical line more or less through its center. Consider the difference between the area enclosed by the figure to the left of the line and the area to the right of the line. Now, if the figure has an exact left–right symmetry, such as the figure shown in N.1b, then clearly the difference will be mathematically zero. But in general, while the difference may be smallish compared to the total area enclosed by the figure, it will, as in A, not be mathematically zero. Conversely, if we are told that the difference is at least 10^{123} times smaller than the area enclosed, we can be almost certain that the figure must be exactly left–right symmetric.

While this geometric analogy hardly does justice to the full glory of the

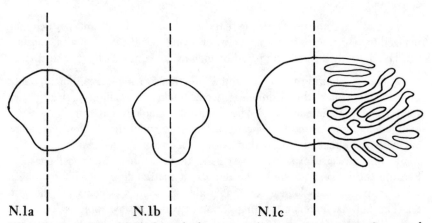

N.1a. *Draw a vertical line through the center of a generic geometric figure and it will divide the figure into two halves roughly equal in area.* **N.1b.** *If the figure is symmetric between left and right, then the area to the left and the area to the right of the line will be mathematically equal.* **N.1c.** *It is possible to have a figure with two halves the same area but vastly different perimeters.*

role played by symmetry in physics, it does convey the spirit of the argument as to why there must be an exact symmetry to account for the fantastic smallness of the cosmological constant. (For a detailed account of the role played by symmetry in physics, see my book *Fearful Symmetry*.)

There does exist a symmetry, named *supersymmetry* by its inventors in tune with this age of hyberbole, which would ensure the smallness of the cosmological constant. Supersymmetry relates one set of particles, such as the photon (members of the set are known as bosons), to another set of particles, such as the electron (members of this second set are known as fermions), just as the left–right symmetry in our geometric analogy relates the left side of the figure to the right side of the figure. The quantum fluctuations of the bosons produce a huge cosmological constant. Similarly, the quantum fluctuations of the fermions. But the supersymmetry relates the bosons and the fermions in precisely such a way that the cosmological constants generated by the bosons and by the fermions cancel each other out, in much the same way that the areas to the left and to the right in our geometric analogy cancel each other out when a symmetry is present.

Wonderful, you say. The cosmological constant paradox is solved; the evil genie is clamped back into the bottle. Alas, the supersymmetry argument does not work.

The trouble is that a symmetry usually has a multitude of consequences. In our analogy, the left–right symmetry not only guarantees that the areas to the left of the center line and to the right are exactly equal, but also that the perimeters of the figure to the left and to the right of the center line are exactly equal. See B. Similarly, supersymmetry requires that the masses of the bosons be exactly the same as that of the fermions. In fact, boson masses and fermion masses are enormously different. Physicists conclude that supersymmetry, if it is present in the final theory of physics at all, must be broken.

The notion of broken symmetry may be understood crudely by supposing that the geometric figure in our analogy is made of rubber. You can start with a left–right symmetric figure, but then you can stretch the figure so it is no longer symmetric. Similarly, physicists start with a symmetric theory and then distort the symmetry. Then, of course, the perimeters of the figure to the left of the dividing line and to the right of the line will no longer be the same. In the same way, physicists can account for the fact that bosons and fermions are observed to have different properties. But—and here is the catch—the cosmological constants generated by the bosons and fermions will no longer cancel each other out, and the cosmological constant comes out huge.

You can't have your cake and eat it, too. In our analogy, if you want the perimeters to the left and to the right of the dividing line to be vastly different, then the areas enclosed by the figure to the left and to the right will no longer be exactly the same. Sure, it is possible, but exceedingly unnatural. See N.1c.

Index

Index

Index

Index

Index